Kaczmarz–Type Methods

for
Solving Linear
and
Nonlinear Equations

求解
线性和非线性方程组的
Kaczmarz 类方法

李维国　邢丽丽　编著
鲍文娣　乔田田

中国石化出版社

内 容 提 要

本书系统介绍了线性和非线性方程组求解中的各种Kaczmarz类算法及其收敛性（率），主要内容包括按顺序的循环Kaczmarz类方法、随机Kaczmarz类方法、稀疏Kaczmarz类方法、非线性Kaczmarz类方法和部分Kaczmarz方法在其他相关问题中的应用。本书内容详实、推导详细、算法分析全面，对于应用Kaczmarz类方法求解大规模数学模型问题具有很好的参考价值。

本书为英文著作，可作为大规模科学与工程计算、机器学习与大数据处理、信号与图像处理等领域本科生、硕士研究生、博士研究生的教学用书，也可作为教学科研一线教师和科研人员的参考书。

图书在版编目（CIP）数据

求解线性和非线性方程组的 Kaczmarz 类方法 = Kaczmarz–Type Methods for Solving Linear and Nonlinear Equations: 英文 / 李维国，邢丽丽，鲍文娣编著 . —北京：中国石化出版社，2022.8
ISBN 978－7－5114－6830－7

Ⅰ.①求…　Ⅱ.①李…②邢…③鲍…　Ⅲ.①线性方程－方程组－数值解－研究－英文②非线性方程－方程组－数值解－研究－英文　Ⅳ.① O241.6

中国版本图书馆 CIP 数据核字（2022）第 142316 号

中国石化出版社出版发行
地址：北京市东城区安定门外大街 58 号
邮编：100011　电话：(010)57512500
发行部电话：(010)57512575
http://www.sinopec-press.com
E-mail：press@sinopec.com
北京捷迅佳彩印刷有限公司印刷
全国各地新华书店经销
*
710×1000 毫米 16 开本 14.25 印张 221 千字
2022 年 8 月第 1 版　2022 年 8 月第 1 次印刷
定价：68.00 元

Preface

The Kaczmarz method was proposed by the Polish mathematician Stefan Kaczmarz in 1937 and the Italian mathematician Gianfranco Cimmino in 1938. It was rediscovered by Gordon R. et al. as an algebraic reconstruction technique in 1970 and reconsidered by Tanabe K. as a projection method in 1971 for solving linear systems of equations. It is an iterative algorithm that has found many applications ranging from computer tomography to digital signal and image processing.

The Kaczmarz method has been widely used for solving mostly over-determined linear systems of equations $Ax = b$ in various fields due to its simple iterative nature with light computation. When the coefficient matrix of systems of equations is with highly coherent rows, the randomized Kaczmarz algorithm is expected to provide faster convergence as it picks a row for each iteration at random, based on a certain probability distribution. After decades of research, especially in the last decade, the relevant methods have been greatly improved and further promoted.

This book surveys the foundational results and more recent works on rates of convergence related to the Kaczmarz algorithms to solve systems of linear and nonlinear equations. The book is divided into five chapters. The first chapter summarizes the conventional Kaczmarz methods of circular projection according to the existing fixed order. In chapter two, various randomized Kaczmarz methods are summarized. Chapter three summarizes the Kaczmarz methods for solving sparse problems, and chapter four focuses on the Kaczmarz methods for solving nonlinear equations. The fifth chapter introduces the application of the Kaczmarz methods in some other problems.

The first, second and fourth chapters of this book contain some of the latest research results of the authors and their research group. Here, we would like to thank our graduate students particularly. They are Guo Junhan, Gao Xingqi, Li Xinkai, Wang Qifeng, Guan Yingjun, Wang Fang, Lv

Zhonglu, Liu Li, Wei Linxiang, Wang Qin and Zhang Feiyu. We also sincerely thank those authors whose research results have been cited by us.

We would also like to thank Bai Hua and Zhao Zixuan of China Petrochemical Press for their warm support and great assistance.

Given the limitations of the authors, it is inevitable that there are some inappropriateness and even mistakes in this book. We sincerely hope that colleagues and readers home and abroad will not hesitate to make their comments.

Li Weiguo, Xing Lili, Bao Wendi and Qiao Tiantian

China University of Petroleum (East China)

July 28, 2022

*This research is supported by the Fundamental Research Funds for the Central Universities (19CX05003A-2) and National Key Research and Development program of China (2019YFC1408400).

Contents

Notation

In this book, if there is no special statement, it is always assumed that $A \in \mathcal{R}^{m \times n}$, $b \in \mathcal{R}^n$. The Euclidean norm of b denoted by $\|b\|_2$, short for $\|b\|$. The spectral norm (Euclidean norm) of A is the quantity $\|A\|_2 = \max\limits_{\|x\|=1} \|Ax\|$, short for $\|A\|$, and the Frobenius norm is $\|A\|_F := \sqrt{\Sigma_{i,j} a_{ij}^2}$. We denote the set of rows of A by $\{a_1^T, \cdots, a_m^T\}$ and the set of columns is denoted $\{A_1, \cdots, A_n\}$. Let $\{\mathcal{I}_1, \mathcal{I}_2, \cdots, \mathcal{I}_s\}$ denote a partition of the $[n] = \{1, 2, \cdots, n\}$ such that, for $i, j = 1, 2, \cdots, s$ and $i \neq j$, $\mathcal{I}_i \neq \emptyset$, $\mathcal{I}_i \bigcap \mathcal{I}_j = \emptyset$, $\bigcup_{i=1}^s \mathcal{I}_i = [n]$. We often use $A_{:, \mathcal{J}}$ (sometimes use $A_{\mathcal{J}}$, if the context can judge) represents the submatrix formed by the column of A composed of the column indexes contained in the index set $\mathcal{J}$, and $A_{\mathcal{I},:}$ (sometimes use $A_{\mathcal{I}}$, if the context can judge) represents a submatrix formed by the row of A composed of the row indexes contained in the index set $\mathcal{I}$ and $A_{\mathcal{I}, \mathcal{J}}$ denotes the submatrix that lies in the rows indexed by $\mathcal{I}$ and the columns indexed by $\mathcal{J}$. We use $|\mathcal{I}|$ to denote the cardinality of a set $\mathcal{I} \subseteq [n]$. Given two symmetric matrices A and B, we use $A \succeq B$ to denote that $A - B$ is positive semidefinite.

For an arbitrary matrix A, denote A^{-1} as the inverse of A, if A is invertible. Otherwise, denote the Moore-Penrose pseudoinverse of A as A^+. $\sigma_1 \geqslant \sigma_2 \geqslant \cdots \geqslant \sigma_r > 0$ as the nonzero singular values of A, and σ_{min}(that is σ_r) and σ_{max}(that is σ_1) as the smallest nonzero singular value and maximum singular value of A respectively. Denote $\lambda_{\min}(A)$, $\lambda_{\max}(A)$ as the smallest nonzero eigenvalue and the maximum eigenvalue of A, respectively.

The relative condition number of A is the quantity $\kappa(A) := \|A\|_2 \|A^{-1}\|_2$ if A is invertible, otherwise, $\kappa(A) := \|A\|_2 \|A^+\|_2$. Related to this is the scaled condition number, defined by $\kappa_F(A) := \|A\|_F \|A^+\|_2$.

If the matrix $A \in \mathcal{R}^{n \times n}$ is symmetric and positive definite, the A-norm of vector $x \in \mathcal{R}^n$ is denoted as $\|x\|_A = \sqrt{x^T A x}$.

Given a nonempty closed convex set $\mathcal{S}$, let $P_{\mathcal{S}}(x)$ be the projection of x onto $\mathcal{S}$: that is, $P_{\mathcal{S}}(x)$ is the vector y that is the optimal solution to $y = \min\limits_{z \in \mathcal{S}} \|x - z\|$. Additionally, define the distance from x to a set $\mathcal{S}$ by

$$d(x, \mathcal{S}) = \min_{z \in \mathcal{S}} \|x - z\| = \|x - P_{\mathcal{S}}(x)\|.$$

The set of coordinate directions in $\mathcal{R}^n$ are denoted by $\{e_1, \cdots, e_n\}$, where e_i is a column vector with the ith entry being 1 and all other entries being zero.

Chapter 1

Kaczmarz Method and Its Variants

1.1 Algebraic Reconstruction Technique

1.1.1 Kaczmarz Method

The Kaczmarz method (proposed by the Polish mathematician Stefan Kaczmarz in 1937[1] (see also its English translation[2]) and the Italian mathematician Gianfranco Cimmino in 1938[3] independently rediscovered by Gordon R. and etc. as the algebraic reconstruction technique in 1970[4], and reconsidered by Tanabe K. as a projection method in 1971[5]) for solving linear systems of equations is an iterative algorithm that has found many applications ranging from computer tomography to digital signal and image processing (see also [6]). Kaczmarz method is also known under the name the algebraic reconstruction technique (ART), which is also a form of alternating projection method.

We will write a consistent linear system of equations as

$$Ax = b, \tag{1.1.1}$$

where $A \in \mathcal{R}^{m \times n}$ and $b \in \mathcal{R}^m$ are given and $x \in \mathcal{R}^n$ is unknown. Then the above system can also be written as

$$\begin{cases} a_{11}x_1 + a_{12}x_2 + \cdots + a_{1n}x_n = b_1, \\ a_{21}x_1 + a_{22}x_2 + \cdots + a_{2n}x_n = b_2, \\ \cdots \quad \cdots \quad \cdots \\ a_{m1}x_1 + a_{m2}x_2 + \cdots + a_{mn}x_n = b_m. \end{cases} \tag{1.1.2}$$

Denote $a_i = (a_{i1}, a_{i2}, \cdots, a_{in})^T$, then we have

$$\langle a_i, \ x \rangle = b_i, i = 1, 2, \cdots, m. \tag{1.1.3}$$

Each equation can be considered as a hyperplane, with a gradient

$$\nabla(\langle a_i, \ x \rangle - b_i) = a_i.$$

Suppose that $x^{(k)}$ is known, then we compute $x^{(k+1)}$ with

$$x^{(k+1)} = x^{(k)} + \beta_i a_i,$$

where

$$\beta_i = \frac{b_i - \langle a_i, \ x^{(k)} \rangle}{\langle a_i, a_i \rangle}$$

can be seen as a step size in the direction a_i, and $\beta_i a_i = \frac{b_i - \langle a_i, x^{(k)} \rangle}{\|a_i\|_2} \frac{a_i}{\|a_i\|_2} = d_i \frac{a_i}{\|a_i\|_2}$, where the geometric meaning of d_i is the directed distance from the point $x^{(k)}$ to the hyperplane $\langle a_i, \ x \rangle = b_i$. Hence

$$x^{(k+1)} = x^{(k)} + \frac{b_i - \langle a_i, \ x^{(k)} \rangle}{\langle a_i, a_i \rangle} a_i = x^{(k)} + \frac{r_i}{\|a_i\|_2} \frac{a_i}{\|a_i\|_2}, \ k = 0, 1, \cdots. \tag{1.1.4}$$

where $r_i = b_i - \langle a_i, \ x^{(k)} \rangle$ is the residual of the ith equation. This is the algebraic reconstruction technique (Kaczmarz method).

This method sweeps through the rows of A in a cyclic manner and involves only a single equation per iteration. The following is the iterative process of this method.

$x^{(0)}$ is given.

Algorithm 1.1.1 Kaczmarz (K) Algorithm

---

procedure $(A, b, x^{(0)}, K)$

    for $k = 0, 1, \cdots, K$ **do**

        Select $i \in [m]$ with $i = k(mod\ m) + 1$

        Set $x^{(k+1)} = x^{(k)} + \frac{b_i - \langle a_i,\ x^{(k)} \rangle}{\|a_i\|_2^2} a_i$

    end for

    Output: $x^{(K+1)}$

end procedure

$k = 0, i = 1$, according to the first equation:

$$x^{(1)} = x^{(0)} + \frac{b_1 - \langle a_1,\ x^{(0)} \rangle}{\|a_1\|^2} a_1;$$

$k = 1, i = 2$, according to the second equation:

$$x^{(2)} = x^{(1)} + \frac{b_2 - \langle a_2,\ x^{(1)} \rangle}{\|a_2\|^2} a_2;$$

$$\vdots$$

$k = m - 1, i = m$, according to the mth equation:

$$x^{(m)} = x^{(m-1)} + \frac{b_m - \langle a_m,\ x^{(m-1)} \rangle}{\|a_m\|^2} a_m;$$

$k = m, i = 1$, according to the first equation:

$$x^{(m+1)} = x^{(m)} + \frac{b_1 - \langle a_1,\ x^{(m)} \rangle}{\|a_1\|^2} a_1;$$

$k = m + 1, i = 2$, according to the second equation:

$$x^{(m+2)} = x^{(m+1)} + \frac{b_2 - \langle a_2,\ x^{(m+1)} \rangle}{\|a_2\|^2} a_2;$$

$$\vdots$$

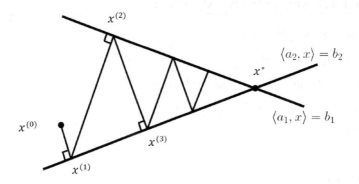

Figure 1.1.1: Kaczmarz method, $m = 2$

Remark 1.1.1. *Kaczmarz proved that (1.1.4) converges to the unique so-lution for square nonsingular matrix A in 1937, but without any attempt to bound the rate of convergence (see [6]).*

Tanabe proved that for any consistent system of equations $Ax = b$, $A \in \mathcal{R}^{m \times n}$, $b \in \mathcal{R}^m$, such that the rows of A are nonzero, and any initial approximation $x^{(0)}$ the sequence generated by $x^{(k+1)} = (P_{H_1} \circ \ldots \circ P_{H_m})(x^{(k)})$ converges to a solution of it, depending on $x^{(0)}$. Where P_{H_i} denotes the projection onto the hyperplane H_i defined by its ith equation (see [6]).

A promotion form of this method in [7] is to add a relaxation factor λ and gets a **relaxed Kaczmarz method**:

$$x^{(k+1)} = x^{(k)} + \lambda_k \frac{b_i - \langle a_i,\ x^{(k)} \rangle}{\|a_i\|^2} a_i.$$

If $\lambda_k \equiv 1$, it is exactly the classical Kaczmarz Method. Nevertheless, the relaxed Kaczmarz method converges even if the linear system $Ax = b$ is overdetermined $(m > n)$ and has no solution. In this case, provided that A has full column rank, the relaxed Kaczmarz method converges to the least squares estimate. This was first observed by Whitney and Meany[8] who proved that the relaxed Kaczmarz method converges provided that the relaxation parameters are within $(0, \frac{2}{\max_i \|a_i\|^2})$ and $\lambda_k \to 0$, when $k \to \infty$.

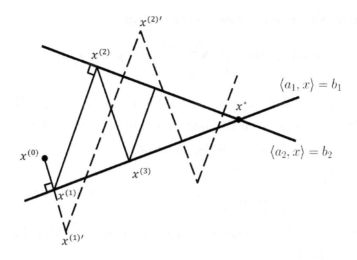

$$\langle a_1, x \rangle = b_1$$

$$\langle a_2, x \rangle = b_2$$

Figure 1.1.2: A promotion form of Kaczmarz method, $m = 2$

In [7], authors proved the following convergence.

Assume that $0 < \lambda_k < 2$. If the system (1.1.1) is consistent then the iteration (1.1.1) converges to a solution x^*, and if $x^{(0)} \in \mathcal{R}(A^T)$ then x^* is the solution of least 2-norm. If the system is inconsistent then every subsequence associated with a_i converges, but not necessarily to a least squares solution.

The $x^{(1)'}, x^{(2)'}, \cdots$ in Figure 1.1.2 is obtained by this method.

Remark 1.1.2. *A symmetric Kaczmarz[9] is an in-order iteration followed by a reverse order iteration. That is*

$$i = 1, 2, \cdots, m, m - 1, \cdots, 1, 2 \cdots.$$

1.1.2 The Simultaneous Iteration Reconstruction Technique

The Simultaneous Iteration Reconstruction Technique (SIRT) which means that all the equations are used at the same time in one iteration was first proposed by Gilbert in 1972[10], which is the correction and improvement of the ART.

Consider using m equations in each iteration

$$x^{(k+1)} = x^{(k)} + \frac{1}{m_i} \sum_{j=1}^{m} \frac{r_j^{(k)}}{\|a_j\|_2^2} a_j, \; k = 0, 1, \cdots, \tag{1.1.5}$$

where m_i is the number of non-zero values in $r_1, r_2, \cdots, r_m$. However, it seems a bit unfair that each component is always divided by m_i. So the SIRT is proposed, which uses all the equations

$$x_l^{(k+1)} = x_l^{(k)} + \frac{1}{m_l} \sum_{j=1}^{m} \frac{r_j^{(k)}}{\|a_j\|_2^2} a_{jl}, l = 1, 2, \cdots, n, \; k = 0, 1, \cdots, \tag{1.1.6}$$

where m_l is the number of non-zero elements of A_l: the lth column of the coefficient matrix A.

Remark 1.1.3. *Here the $\|a_j\|_2$, $j = 1, \cdots, m$; m_l, $l = 1, \cdots, m$ can be all calculated before the iteration.*

Remark 1.1.4. *A promotion form of this method*

$$x_l^{(k+1)} = x_l^{(k)} + \frac{\omega}{\sum\limits_{j=1}^{m} |a_{jl}|^\alpha} \sum_{j=1}^{m} \frac{r_j^{(k)}}{\sum\limits_{l=1}^{n} |a_{jl}|^{2-\alpha}} a_{jl}, l = 1, 2, \cdots, n, \; k = 0, 1, \cdots.$$

$$\tag{1.1.7}$$

Especially, when $\alpha = 0, \omega = 1$, same to (1.1.6), it was given by Dines and Lyttle (1979)[11]; When $\alpha = 1, \omega = 1$, it was the Hager method (1985)[12].

Remark 1.1.5. *SIRT can be also written in the general matrix form [9]:*

$$x^{(k+1)} = x^{(k)} + \lambda_k T A^T M (b - A x^{(k)}), \; k = 0, 1, 2, \cdots \tag{1.1.8}$$

where the matrices M and T are symmetric positive definite. Different methods depend on the choice of these matrices. The following theorem regarding convergence summarizes the results from [13]-[14].

Theorem 1.1.1. *The iterates of the form (1.1.8) converge to a solution x^* of $\min\limits_{x} \|Ax - b\|_M$ if and only if*

$$0 < \epsilon \leqslant \lambda_k \leqslant \frac{2 - \epsilon}{\rho(T A^T M A)},$$

where ϵ is an arbitrarily small but fixed constant. In addition $x^{(0)} \in \mathcal{R}(TA^T)$ then x^* is the unique solution of minimum T^{-1}-norm (minimum 2-norm if $T = I$).

SIRT contains several methods are as follows.

Landweber's method. The classical Landweber method takes the form:

$$x^{(k+1)} = x^{(k)} + \lambda_k A^T(b - Ax^{(k)}), \ k = 0, 1, 2, \cdots, \tag{1.1.9}$$

which corresponds to the setting $M = I$ and $T = I$ in (1.1.8).

Cimmino's method. It is often presented in a variant based on projections. The next iterate $x^{(k+1)}$ is the average of the projections of the previous iterate $x^{(k)}$ on all the hyperplanes H_i for $i = 1, 2, \cdots, m$:

$$x^{(k+1)} = x^{(k)} + \frac{1}{m} \sum_{j=1}^{m} \frac{r_j^{(k)}}{\|a_j\|^2} a_j, \ k = 0, 1, \cdots. \tag{1.1.10}$$

The general version of Cimmino's method includes a relaxation parameter λ_k as well as weights w_j:

$$x^{(k+1)} = x^{(k)} + \frac{\lambda_k}{m} \sum_{j=1}^{m} \frac{w_j r_j^{(k)}}{\|a_j\|^2} a_j, \ k = 0, 1, \cdots. \tag{1.1.11}$$

With the use of matrix notation the general Cimmino's method takes the form of (1.1.8) with $M = D$ and $T = I$, where we have defined $D = \frac{1}{m} diag\left(\frac{w_j}{\|a_j\|_2^2}\right)$.

Component averaging method (CAV). Cimmino's original method uses equal weighting of the contributions from the projections, which seems fair when A is a dense matrix. Component Averaging (CAV) is as an extension of Cimmino's method which incorporates information about the sparsity of A (if any), in a heuristic way. Let s_j denote the number of nonzero elements of A_j, column j of A: $s_j = NNZ(A_j), \ j = 1, 2, \cdots, n$, and define the diagonal matrix $S = diag(s_1, \cdots, s_n)$ and the norm $\|a_i\|_S^2 =$

$\langle a_i, S a_i \rangle = \sum_{j=1}^{n} a_{ij}^2 s_j$, for $i = 1, \cdots, m$. Then the CAV algorithm takes the form:

$$x^{(k+1)} = x^{(k)} + \lambda_k \sum_{j=1}^{m} \frac{w_j r_j^{(k)}}{\|a_j\|_S^2} a_j, \ k = 0, 1, 2, \cdots, \tag{1.1.12}$$

and when A is dense then $S = mI$ and we get Cimmino's method. The CAV algorithm thus takes the matrix form (1.1.8) with $M = D_S$ and $T = I$, where we have defined $D_S = diag\left(\frac{w_j}{\|a_j\|_S^2}\right)$.

Diagonally Relaxed Orthogonal Projections (DROP). This method is another extension of Cimmino's original method, in which the factors s_j, same as CAV, are incorporated in a different manner, namely, by computing the next iterate as

$$x^{(k+1)} = x^{(k)} + \lambda_k S^{-1} \sum_{j=1}^{m} \frac{w_j r_j^{(k)}}{\|a_j\|_2^2} a_j. \ k = 0, 1, \cdots. \tag{1.1.13}$$

The DROP method thus has the form (1.1.8) with $T = S^{-1}$ and $M = mD$, with $D = \frac{1}{m} diag\left(\frac{w_j}{\|a_j\|_2^2}\right)$. Again we obtain Cimmino's method when A is dense in which case $S^{-1} = m^{-1}I$. It is shown in [15] that $\rho(S^{-1}A^T M A) \leqslant \max_i\{w_i\}$, which means that convergence is guaranteed if $\lambda_k \leqslant \frac{2-\epsilon}{\max_i\{w_i\}}$ where ϵ is an arbitrarily small but fixed constant. Numerical experiments show that it is worthwhile to use the larger upper bound $\frac{2}{\rho(S^{-1}A^T M A)}$ for λ_k, instead of the easily computed bound $\frac{2}{\max_i\{w_i\}}$.

Simultaneous Algebraic Reconstruction Technique (SART). This method was originally developed in the ART setting[16], but it can also be written and implemented in the SIRT form (1.1.8) and we therefore categorize it as an SIRT method. It is written in the following matrix form:

$$x^{(k+1)} = x^{(k)} + \lambda_k D_c^{-1} A^T D_r^{-1} (b - A x^{(k)}), \ k = 0, 1, \cdots, \tag{1.1.14}$$

where the diagonal matrices D_r and D_c are defined in terms of the row and the column sums: $D_r = diag(\|a_i\|_1)$, $D_c = diag(\|A_j\|_1)$. We do not include weights in this method. The convergence for SART was independently established in [13] and [17], where it is shown that $\rho(D_c^{-1} A^T D_r^{-1} A) = 1$ and that convergence therefore is guaranteed for $0 < \lambda_k < 2$.

In [18], several methods for derivation of SART and connections between SART and other methods are provided. Using these connections, the convergences of SART in different ways are proved.

A MATLAB package with implementations of several algebraic iterative reconstruction methods for discretizations of inverse problems are presented in [9].

1.2 Coordinate Descent Method

Consider the **coordinate descent method** of linear systems.

1. Let $A \in \mathcal{R}^{n \times n}$ be a symmetric positive-definite matrix, then the linear system $Ax = b$ has a unique solution $x^* = A^{-1} b$. We consider the equivalent problem of minimizing the strictly convex quadratic function

$$f(x) = \frac{1}{2} x^T A x - b^T x,$$

and note the standard relationship

$$f(x) - f(x^*) = \frac{1}{2} \| x - x^* \|_A^2. \tag{1.2.1}$$

Suppose our current iterate is $x^{(k)}$ and we obtain a new iterate $x^{(k+1)}$ by performing an exact line search in the nonzero direction d, that is, $x^{(k+1)}$ is the solution to $\min_{t \in \mathcal{R}} f(x + td)$. This gives us

$$x^{(k+1)} = x^{(k)} + \frac{(b - Ax^{(k)})^T d}{d^T A d} d \tag{1.2.2}$$

and

$$f(x^{(k+1)}) - f(x^*) = \frac{1}{2} \| x^{(k+1)} - x^* \|_A^2 = \frac{1}{2} \| x^{(k)} - x^* \|_A^2 - \frac{((b - Ax^{(k)})^T d)^2}{2 d^T A d}.$$

One natural choice of a set of easily-computable search directions is to choose d from the set of coordinate directions, $\{e_1, \cdots, e_n\}$. Note that, when using search direction e_i, we can compute the new point

$$x^{(k+1)} = x^{(k)} + \frac{b_i - \langle a_i, x^{(k)} \rangle}{a_{ii}} e_i \qquad (1.2.3)$$

using only $2n + 2$ arithmetic operations. If the search direction is chosen at each iteration by successively cycling through the set of coordinate directions, then the algorithm is known to be linearly convergent but with a rate not easily expressible in terms of typical matrix quantities (see [19]). However, by choosing a coordinate direction as a search direction randomly according to an appropriate probability distribution, the convergence rate in terms of the relative condition number can be obtained. This is expressed in the following chapter.

2. If $A \in \mathcal{R}^{n \times n}$ is a **symmetric positive-semidefinite matrix, and** $a_{ii} > 0$ for all $i \in [n]$, the iterative method (1.2.3) is still available. The expected convergence rate is expressed in the following chapter.

3. If $A \in \mathcal{R}^{m \times n}$ is a **general matrix, and** $\|A_i\| \neq 0$ for all $i \in [n]$, it can be considered for the more general problem of finding a solution to a linear system $Ax = b$. More generally, since the system might be inconsistent, the least squares solution by minimizing the function $\|b - Ax\|^2$ is sought. The minimizers are exactly the solutions of the positive semidefinite system $A^T Ax = A^T b$, to which we could easily apply the previous algorithm. Now we get an iterative formula as follows:

$$\begin{aligned} x^{(k+1)} &= x^{(k)} + \frac{\langle A^T b - A^T A x^{(k)}, e_i \rangle}{e_i^T A^T A e_i} e_i \\ &= x^{(k)} + \frac{\langle b - A x^{(k)}, A e_i \rangle}{e_i^T A^T A e_i} e_i \\ &= x^{(k)} + \frac{\langle r^{(k)}, A_i \rangle}{\|A_i\|^2} e_i. \end{aligned} \qquad (1.2.4)$$

As usual, we avoid computing the new matrix $A^T A$ explicitly. Instead, we can proceed as following coordinate descent (CD) method which is also called Gauss-Seidel method [20].

By the coordinate descent nature of this algorithm, once one have computed the initial residual $r^{(0)}$ and column norms $\{\|A_i\|^2\}_{i=1}^n$, we can perfor-

---

Algorithm 1.2.1 Coordinate Descent (CD) Algorithm

---

procedure $(A,\ b,\ x^{(0)},\ K,\ r^{(0)} = b - Ax^{(0)})$

  for $k = 0, 1, \cdots, K$ **do**

    Select $i \in [n]$ with $i = k(mod\ n) + 1$

    Compute $\alpha_k = \frac{\langle A_i, r^{(k)} \rangle}{\|A_i\|^2}$,

    Compute $x^{(k+1)} = x^{(k)} + \alpha_k e_i$ and $r^{(k+1)} = r^{(k)} - \alpha_k A_i$

  end for

  Output: $x^{(K+1)}$

end procedure

---

m each iteration in $O(m)$ time, just as in the positive-definite case. Specifically, this new iteration takes $4m + 1$ arithmetic operations, compared with $2n + 2$ for the positive-definite case.

We would expect that Algorithm 1.2.1 converges to a least squares solution, even in the case where the underlying system is inconsistent. The next chapter shows that this is, in fact, the case.

Remark 1.2.1. *In fact,* **Kaczmarz method can also be regarded as a CD method.** *If $A \in \mathcal{R}^{m \times n}$ is a general matrix, and $\|a_i\| \neq 0$ for all $i \in [m]$, we consider the following regularizing linear system*

$$AA^T y = b, \quad x = A^T y.$$

Obviously, AA^T is a positive semidefinite matrix. With the previous algorithm, we can obtain an iterative formula as follows:

$$
\begin{aligned}
y^{(k+1)} &= y^{(k)} + \frac{\langle b - AA^T y^{(k)}, e_i \rangle}{e_i^T AA^T e_i} e_i \\
&= y^{(k)} + \frac{\langle b - Ax^{(k)}, e_i \rangle}{\|a_i\|^2} e_i \\
&= y^{(k)} + \frac{r_i^{(k)}}{\|a_i\|^2} e_i.
\end{aligned}
\tag{1.2.5}
$$

Multiply both sides by A^T, we get

$$x^{(k+1)} = x^{(k)} + \frac{r_i^{(k)}}{\|a_i\|^2} a_i.$$

This is the Kaczmarz method as (1.1.4).

Remark 1.2.2. *The RK update (1.1.4) is equivalent to one step of coordinate descent applied to the dual problem*

$$\min_{y \in \mathcal{R}^m} \{\frac{1}{2} y^T A A^T y - y^T b\},$$

(specifically, a negative gradient step in the ith component of y with steplength $1/\|a_i\|^2$), where the primal variable x and dual y are related through $x = A^T y$.

1.3   Maximal Correction Kaczmarz Method

1.3.1   MCK For Linear Problem

There are at least two greedy selection rules: the maximum residual (MR) rule and the maximum distance (MD) rule, respectively:

$$i_k = \arg\max_i |\langle a_i, x \rangle - b_i| \ \text{(MR)}; \quad i_k = \arg\max_i |\langle a_i, x \rangle - b_i| / \|a_i\|_2 \ \text{(MD)}$$

$$(1.3.1)$$

where i_k is the row index that should be selected at the *kth* iteration.

As a result of this idea, the maximal correction Kaczmarz (MCK) method is proposed, which gives a new rule to choose one row of A with maximal correction to the distances of current iterative point to the hyperplanes. Here is the proposed algorithm.

The convergence analyses are as follow.

Lemma 1.3.1. *If $\alpha_1, \alpha_2, \cdots, \alpha_m$ and $\beta_1, \beta_2, \cdots, \beta_m$ are two nonnegative sequences such that $\sum\limits_{k=1}^{m} \alpha_k = 1$, then the following inequality is true*

$$\sum_{k=1}^{m} \alpha_k \beta_k \leqslant \max_{1 \leqslant j \leqslant m} \beta_j. \qquad (1.3.2)$$

If $\alpha_i \neq 0, i = 1, 2, \cdots, m$, the equality holds if and only if $\beta_1 = \cdots = \beta_m$.

Algorithm 1.3.1 Maximal Correction Kaczmarz (MCK) Algorithm

---

procedure $(A,\ b,\ x^{(0)},\ K,\ r = b - Ax^{(0)})$

    for $k = 0, 1, \cdots, K$ **do**

        Select $q \in [m]$ with $[*, q] = \max\limits_{i} \frac{|r_i|}{\|a_i\|}$

        Set $x^{(k+1)} = x^{(k)} + \frac{r_q}{\|a_q\|^2} a_q$

        Compute $r = b - Ax^{(k+1)}$

    end for

    Output: $x^{(K+1)}$

end procedure

Because the proof is too simple, we omit it.

Theorem 1.3.1. *Let x^* be the least-norm solution of (1.1.1) and $x^{(0)} \in \mathcal{R}(A^T)$, then the sequence $\{x^{(k)}\}$ generated from the MCK linearly converges to x^*.*

Proof. From the inequality (1.3.2), we have

$$\sum_{j=1}^{m} \frac{\|a_j\|^2}{\|A\|_F^2} \left| \left\langle x^{(k)} - x^*, \frac{a_j}{\|a_j\|} \right\rangle \right|^2 \leqslant \left| \left\langle x^{(k)} - x^*, \frac{a_q}{\|a_q\|} \right\rangle \right|^2 = \frac{\left| r_q^{(k)} \right|^2}{\|a_q\|^2}, \quad (1.3.3)$$

where the index $q = \arg\max\limits_{1 \leqslant j \leqslant m} \left| \left\langle x^{(k)} - x^*, \frac{a_j}{\|a_j\|} \right\rangle \right| = \arg\max\limits_{1 \leqslant j \leqslant m} \frac{\left| r_j^{(k)} \right|}{\|a_j\|}$. It is easy to get the following inequality by using extreme value inequality

$$\sum_{j=1}^{m} \frac{\|a_j\|^2}{\|A\|_F^2} \left| \left\langle x^{(k)} - x^*, \frac{a_j}{\|a_j\|} \right\rangle \right|^2 = \frac{1}{\|A\|_F^2} \sum_{j=1}^{m} |\langle x^{(k)} - x^*, a_j \rangle|^2$$

$$= \frac{1}{\|A\|_F^2} \|A(x^{(k)} - x^*)\|^2$$

$$\geqslant \frac{\|x^{(k)} - x^*\|^2}{\|A\|_F^2 \|A^+\|^2} = \frac{1}{\kappa_F^2(A)} \|x^{(k)} - x^*\|^2.$$

$$(1.3.4)$$

With (1.3.3) and (1.3.4), we have

$$\left| \left\langle x^{(k)} - x^*, \frac{a_q}{\|a_q\|} \right\rangle \right|^2 \geqslant \kappa_F^{-2}(A) \|x^{(k)} - x^*\|^2. \quad (1.3.5)$$

By the orthogonality of $x^{(k)} - x^{(k+1)}$ and $x^{(k+1)} - x^*$, it holds

$$
\begin{aligned}
\|x^{(k+1)} - x^*\|^2 &= \|x^{(k)} - x^*\|^2 - \|x^{(k+1)} - x^{(k)}\|^2 \\
&= \|x^{(k)} - x^*\|^2 - \frac{\left|r_q^{(k)}\right|^2}{\|a_q\|^2} \\
&= \|x^{(k)} - x^*\|^2 - \left|\left\langle x^{(k)} - x^*, \frac{a_q}{\|a_q\|}\right\rangle\right|^2 \\
&\leqslant \left(1 - \kappa_F^{-2}(A)\right) \|x^{(k)} - x^*\|^2.
\end{aligned}
\tag{1.3.6}
$$

Thus we obtain the linear convergence rate of the MCK.   □

Remark 1.3.1. *The MCK algorithm can be further simplified if $B = AA^T$ can be calculated in advance. At this time, the step 5 can be calculated as follows: $r = r - \frac{r_q}{\|a_q\|^2} B(:, q)$.*

Remark 1.3.2. *Can we find a faster convergence direction than the orthogonal projection direction (see Figure 1.1.1)? How to find the direction from $x^{(k)}$ to x^* directly with four points that coexist with a circle?*

1.3.2   Finite Termination of MCK Method

Obviously, the Maximal Correction Kaczmarz algorithm has local optimality. As a special case, when the rows of matrix A are orthogonal for each other, the algorithm will converge to the solution x^* with at most m steps.

The matrix A is row orthogonal, without losing generality. Assume that

$$
x^* - x^{(0)} = \sum_{j=1}^{m} \mu_j \frac{a_j}{\|a_j\|}.
$$

We prove this conclusion by induction. When $m = 2$,

Step 1: Assume that $\mu_1 \geqslant \mu_2$, then the MCK algorithm chooses $q = 1$ and

$$
x^{(1)} = x^{(0)} + \frac{r_1}{\|a_1\|} \cdot \frac{a_1}{\|a_1\|} \equiv x^{(0)} + \mu_1 \frac{a_1}{\|a_1\|},
$$

where $\mu_1 = \langle x^* - x^{(0)}, \frac{a_1}{\|a_1\|} \rangle$. So we have

$$x^* - x^{(1)} = x^* - x^{(0)} - \mu_1 \frac{a_1}{\|a_1\|} = \mu_2 \frac{a_2}{\|a_2\|}.$$

Step 2: Here $q = 2$,

$$x^{(2)} = x^{(1)} + \frac{r_2}{\|a_2\|} \cdot \frac{a_2}{\|a_2\|} \equiv x^{(0)} + \mu_1 \frac{a_1}{\|a_1\|} + \mu_2 \frac{a_2}{\|a_2\|} = x^*.$$

Assume that the conclusion holds when $m = k - 1$. When $m = k$:

Step 1: Assume that $\mu_1 = \max_{1 \leqslant i \leqslant m} \mu_j$. The MCK method chooses $q = 1$ and

$$x^{(1)} = x^{(0)} + \frac{r_1}{\|a_1\|} \cdot \frac{a_1}{\|a_1\|} \equiv x^{(0)} + \mu_1 \frac{a_1}{\|a_1\|}.$$

Now,

$$x^* - x^{(1)} = \sum_{j=2}^{m} \mu_j \frac{a_j}{\|a_j\|}.$$

So the conclusion is established from the inductive hypothesis.

Remark 1.3.3. *Are there any methods and conclusions similar to MCK method in nonlinear problems?*

1.4   Kaczmarz Method with Oblique Projection

In Remark 1.3.2, a question is raised: if we have computed $x^{(k)}$, how to find the next iteration $x^{(k+1)}$ which is on the intersection of two hyperplanes? Here we give a simple strategy to find a next iteration $x^{(k+1)}$ that can converge to $\tilde{x}$ much quickly, where $\tilde{x}$ is a solution of the linear system (1.1.1).

Our Kaczmarz method with oblique projection (KO) is described as follows (refer to Figure 1.4.1):

Consider twice successive projections of two adjacent hyperplanes. Assume that $x^{(k)}$ is the kth iteration of solving the systems of equations (1.1.1), and $x^{(k)}$ is on the hyperplane $\langle a_{i_k}, x \rangle = b_{i_k}$. By using an orthogonal projection from point $x^{(k)}$ to the hyperplane $\langle a_{i_{k+1}}, x \rangle = b_{i_{k+1}}$, we get the projection point $y^{(k)}$. And then with an orthogonal projection from point

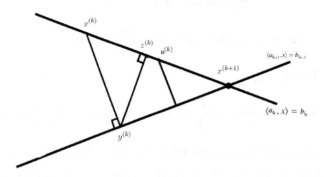

Figure 1.4.1: KO, $m = 2$

$y^{(k)}$ to the hyperplane $\langle a_{i_k}, x \rangle = b_{i_k}$, the projection point $z^{(k)}$ is given. Let line L pass through point $x^{(k)}$ and along direction $w^{(i_k)} = z^{(k)} - x^{(k)}$, then the intersection of L and the hyperplane $\langle a_{i_{k+1}}, x \rangle = b_{i_{k+1}}$ are chosen as the next iteration point $x^{(k+1)}$ (denote $x^{(k+1)}$ as the oblique projection point of $x^{(k)}$ along $w^{(i_k)}$ to the hyperplane $\langle a_{i_{k+1}}, x \rangle = b_{i_{k+1}}$).

Assume that the middle point of the line segment which connects the point $x^{(k)}$ to the point $x^{(k+1)}$ is $u^{(k)}$, then

$$u^{(k)} = x^{(k)} + t_k w^{(i_k)} \quad and \quad x^{(k+1)} = x^{(k)} + 2t_k w^{(i_k)}. \tag{1.4.1}$$

If the directed distance from the point $x^{(k)}$ to the hyperplane $\langle a_{i_{k+1}}, x \rangle = b_{i_{k+1}}$ is denoted as d_k

$$d_k = \frac{b_{i_{k+1}} - \langle a_{i_{k+1}}, x^{(k)} \rangle}{\|a_{i_{k+1}}\|}, \tag{1.4.2}$$

then the directed distance $\frac{1}{2}d_k$ from the point $u^{(k)}$ to the hyperplane $\langle a_{i_{k+1}}, x \rangle = b_{i_{k+1}}$ is determined by

$$\frac{1}{2}d_k = \frac{b_{i_{k+1}} - \langle a_{i_{k+1}}, u^{(k)} \rangle}{\|a_{i_{k+1}}\|}. \tag{1.4.3}$$

With the use of the first equality of (1.4.1), we have

$$t_k = \frac{b_{i_{k+1}} - \langle a_{i_{k+1}}, x^{(k)} \rangle}{2\langle a_{i_{k+1}}, w^{(i_k)} \rangle}, \tag{1.4.4}$$

if $\langle a_{i_{k+1}}, w^{(i_k)} \rangle \neq 0$ ($\langle a_{i_{k+1}}, w^{(i_k)} \rangle = 0$ corresponds to the hyperplane $\langle a_{i_k}, x \rangle = b_{i_k}$ parallel to the hyperplane $\langle a_{i_{k+1}}, x \rangle = b_{i_{k+1}}$). By the second equality of (1.4.1), we get

$$x^{(k+1)} = x^{(k)} + \frac{b_{i_{k+1}} - \langle a_{i_{k+1}}, x^{(k)} \rangle}{\langle a_{i_{k+1}}, w^{(i_k)} \rangle} w^{(i_k)}, \tag{1.4.5}$$

where $w^{(i_k)} = z^{(k)} - x^{(k)} = \frac{b_{i_{k+1}} - \langle a_{i_{k+1}}, x^{(k)} \rangle}{\|a_{i_{k+1}}\|^2} a_{i_{k+1}} + \frac{b_{i_k} - \langle a_{i_k}, y^{(k)} \rangle}{\|a_{i_k}\|^2} a_{i_k}$ (The orthogonal projection of Kaczmarz method is used directly). Set that $w^{(i_k)} = \lambda(a_{i_{k+1}} + \mu a_{i_k})$ (for convenience, the factor λ could be ignored). According to the orthogonality of $w^{(i_k)}$ and a_{i_k}, we get $\mu = -\frac{\langle a_{i_k}, a_{i_{k+1}} \rangle}{\langle a_{i_k}, a_{i_k} \rangle}$ and

$$w^{(i_k)} = a_{i_{k+1}} - \frac{\langle a_{i_k}, a_{i_{k+1}} \rangle}{\langle a_{i_k}, a_{i_k} \rangle} a_{i_k}. \tag{1.4.6}$$

The algorithm is described as in Algorithm 1.4.1. Without losing generality, we assume that all rows of A are not zero vectors.

Assume the system (1.1.1) is consistent, then it must be $b_{i_k} = \lambda b_{i_{k+1}}$ if the two rows of the coefficient matrix A have relation $a_{i_k} = \lambda a_{i_{k+1}}$. In this case, the two hyperplanes $\langle a_{i_k}, x \rangle = b_{i_k}$ and $\langle a_{i_{k+1}}, x \rangle = b_{i_{k+1}}$ are coincident, and we can eliminate one of them without affecting the solution of the equations. So in the following proof, we always assume that $0 < \theta_{i_{k+1}} \leqslant \pi/2$, here $\theta_{i_{k+1}}$ is the angle between any two hyperplanes $\langle a_{i_k}, x \rangle = b_{i_k}$ and $\langle a_{i_{k+1}}, x \rangle = b_{i_{k+1}}$. In Algorithm 1.4.1,

$$h_{i_k} = \|w^{(i_k)}\|^2 = \frac{1}{\|a_{i_k}\|^2} \|a_{i_{k+1}}\|^2 \|a_{i_k}\|^2 \left(1 - \cos^2(\theta_{i_k})\right) = \|a_{i_{k+1}}\|^2 \sin^2(\theta_{i_{k+1}}).$$

Thus $h_{i_k} \geqslant \epsilon > 0$ ensures $\theta_{i_{k+1}} > 0$ because $h_{i_k} = \|a_{i_{k+1}}\|^2 \sin^2(\theta_{i_{k+1}})$.

Before giving the proof of the convergence of the KO algorithm, we first restate the KO algorithm as the following process. For $x^{(0)} \in \mathcal{R}^n$ as an

---

Algorithm 1.4.1 Kaczmarz Method with Oblique Projection (KO)

---

procedure $(A,\ b,\ x^{(0)},\ K,\ \varepsilon > 0)$

For $i = 1:m$, $M(i) = \|a_i\|^2$

Compute $x^{(1)} = x^{(0)} + \frac{b_1 - \langle a_1, x^{(0)} \rangle}{M(1)} a_1$ and set $i_{k+1} = 1$

for $k = 1, 2, \cdots, K$ **do**

Set $i_k = i_{k+1}$ and choose a new i_{k+1}: $i_{k+1} = mod(k, m) + 1$

Compute $D_{i_k} = \langle a_{i_k}, a_{i_{k+1}} \rangle$ and $r_{i_{k+1}}^{(k)} = b_{i_{k+1}} - \langle a_{i_{k+1}}, x^{(k)} \rangle$

Compute $w^{(i_k)} = a_{i_{k+1}} - \frac{D_{i_k}}{M(i_k)} a_{i_k}$ and $h_{i_k}(= \|w^{(i_k)}\|^2) = M(i_{k+1}) - \frac{D_{i_k}}{M(i_k)} D_{i_k}$

if $h_{i_k} > \varepsilon$ **then**

$\alpha_{i_k}^{(k)} = \frac{r_{i_{k+1}}^{(k)}}{h_{i_k}}$ and $x^{(k+1)} = x^{(k)} + \alpha_{i_k}^{(k)} w^{(i_k)}$

end if

end for

Output: $x^{(K+1)}$

end procedure

---

initial approximation we successively define $x^{(0,0)}, x^{(0,1)}, \cdots, x^{(0,m)} \in \mathcal{R}^n$ by

$$
\begin{cases}
x^{(0,0)} = x^{(0)} + \frac{b_1 - \langle a_1, x^{(0)} \rangle}{\|a_1\|^2} a_1, \\
x^{(0,1)} = x^{(0,0)} + \frac{b_2 - \langle a_2, x^{(0,0)} \rangle}{\|w^{(1)}\|^2} w^{(1)}, \\
x^{(0,2)} = x^{(0,1)} + \frac{b_3 - \langle a_3, x^{(0,1)} \rangle}{\|w^{(2)}\|^2} w^{(2)}, \\
\cdots\cdots\cdots\cdots\cdots\cdots\cdots \\
x^{(0,m-1)} = x^{(0,m-2)} + \frac{b_m - \langle a_m, x^{(0,m-2)} \rangle}{\|w^{(m-1)}\|^2} w^{(m-1)}, \\
x^{(0,m)} = x^{(0,m-1)} + \frac{b_1 - \langle a_1, x^{(0,m-1)} \rangle}{\|w^{(m)}\|^2} w^{(m)},
\end{cases}
\tag{1.4.7}
$$

where

$$
w^{(i)} = a_{i+1} - \frac{\langle a_{i+1}, a_i \rangle}{\langle a_i, a_i \rangle} a_i, \ i = 1, \cdots, m-1, \ w^{(m)} = a_1 - \frac{\langle a_1, a_m \rangle}{\langle a_m, a_m \rangle} a_m.
\tag{1.4.8}
$$

For convenience, we denote $a_{m+1} \equiv a_1$, $b_{m+1} \equiv b_1$. Then, for an arbitrary $p \geqslant 0$ and a given approximation $x^{(p,m)} \in \mathcal{R}^n$, we successively construct the new ones $x^{(p+1,1)}, x^{(p+1,2)}, \cdots, x^{(p+1,m)} \in \mathcal{R}^n$ by

$$
\begin{cases}
for \ \ i = 1 : m \\
x^{(p+1,i)} = x^{(p+1,i-1)} + \frac{b_{i+1} - \langle a_{i+1}, x^{(p+1,i-1)} \rangle}{\|w^{(i)}\|^2} w^{(i)}, \\
end
\end{cases}
\tag{1.4.9}
$$

with the notational convention

$$
x^{(p+1,0)} = x^{(p,m)}.
\tag{1.4.10}
$$

Obviously, $x^{(k+1)} = x^{(p,i)}$, if $k = p \cdot m + i$, $0 \leqslant i < m$. The convergence of KO is provided as follows.

Theorem 1.4.1. *Let $x^{(0)} \in \mathcal{R}^n$ be an arbitrary initial approximation. $\tilde{x}$ is a solution of (1.1.1) such that $P_{\mathcal{N}(A)}(\tilde{x}) = P_{\mathcal{N}(A)}(x^{(0)})$, and the sequence $\{x^{(k)}\}_{k=1}^{\infty}$ is generated with the KO algorithm. Then,*

$$
\lim_{k \to \infty} x^{(k)} = \tilde{x}.
\tag{1.4.11}
$$

In addition, if $x^{(0)} \in \mathcal{R}(A^T)$, then $\{x^{(k)}\}$ converges to the least-norm solution of (1.1.1), i.e.,

$$
\lim_{k \to \infty} x^{(k)} = x^*.
$$

Proof. According to (1.4.7)-(1.4.10), we obtain the sequence of approximations (from top to bottom and left to right, and also by using the notational
convention (1.4.10))

$$
\begin{cases}
x^{(0)}, x^{(0,0)} \\
x^{(0,1)}, x^{(0,2)}, \cdots, x^{(0,m)} = x^{(1,0)} \\
x^{(1,1)}, x^{(1,2)}, \cdots, x^{(1,m)} = x^{(2,0)} \\
\cdots\cdots\cdots\cdots\cdots\cdots\cdots\cdots \\
x^{(p,1)}, x^{(p,2)}, \cdots, x^{(p,m)} = x^{(p+1,0)} \\
\cdots\cdots\cdots\cdots\cdots\cdots\cdots\cdots
\end{cases}
\tag{1.4.12}
$$

We define the numbers

$$
r^{(p,i)} = b_{i+1} - \langle a_{i+1}, x^{(p,i-1)} \rangle, \quad i = 1, 2, \cdots, m, \ \forall p \geqslant 0.
\tag{1.4.13}
$$

From (1.4.7)-(1.4.13), it results in

$$
x^{(p,i)} = x^{(p,i-1)} + \frac{r^{(p,i)}}{\|w^{(i)}\|^2} w^{(i)}, \quad p \geqslant 0, \ i = 1, 2, \cdots, m.
\tag{1.4.14}
$$

Based on the description of the KO algorithm (refer to Fig. 1.4.2), $x^{(p,i)}$ is
the oblique projection point of $x^{(p,i-1)}$ along $w^{(i)}$ to the hyperplane $\langle a_{i+1}, x \rangle = b_{i+1}$. So the points $x^{(p,i)}$ and $\tilde{x}$ are on the hyperplane $\langle a_{i+1}, x \rangle = b_{i+1}$, then
$\langle a_{i+1}, x^{(p,i)} - \tilde{x} \rangle = 0$. In addition, the points $x^{(p,i)}$ and $\tilde{x}$ are also on the
hyperplane $\langle a_i, x \rangle = b_i$, therefore, $\langle a_i, x^{(p,i)} - \tilde{x} \rangle = 0$. According to the
definition of $w^{(i)}$, $\langle w^{(i)}, x^{(p,i)} - \tilde{x} \rangle = 0$, that is,

$$
\langle x^{(p,i-1)} - x^{(p,i)}, x^{(p,i)} - \tilde{x} \rangle = 0.
\tag{1.4.15}
$$

Therefore, from $x^{(p,i-1)} - \tilde{x} = x^{(p,i-1)} - x^{(p,i)} + x^{(p,i)} - \tilde{x}$, it is easy to see

$$
\|x^{(p,i-1)} - \tilde{x}\|^2 = \|x^{(p,i-1)} - x^{(p,i)}\|^2 + \|x^{(p,i)} - \tilde{x}\|^2.
$$

From (1.4.14), we get

$$
\|x^{(p,i-1)} - \tilde{x}\|^2 = \|x^{(p,i)} - \tilde{x}\|^2 + \frac{|r^{(p,i)}|^2}{\|w^{(i)}\|^2}.
\tag{1.4.16}
$$

Obviously, the sequence $\{\|x^{(p,i)} - \tilde{x}\|\}_{p=0,i=1}^{\infty,m}$, i.e., $\{\|x^{(k+1)} - \tilde{x}\|\}_{k=0}^{\infty}$ is a monotonically decreasing sequence with lower bounds. There exists a $\alpha \geqslant 0$ such that

$$\lim_{p \to \infty} \|x^{(p,i)} - \tilde{x}\| = \alpha \geqslant 0, \quad \forall \, i = 1, 2, \cdots, m. \tag{1.4.17}$$

Thus, from (1.4.16) and because i is arbitrary we get

$$\lim_{p \to \infty} r^{(p,i)} = 0, \quad \forall \, i = 1, 2, \cdots, m. \tag{1.4.18}$$

Because the sequence $\{\|x^{(p,i)} - \tilde{x}\|\}_{p=0,i=1}^{\infty,m}$ is bounded, we obtain

$$\|x^{(p,i)}\| \leqslant \|\tilde{x}\| + \|x^{(p,i)} - \tilde{x}\| \leqslant \|\tilde{x}\| + \|x^{(0,1)} - \tilde{x}\|, \ \forall p \geqslant 0. \tag{1.4.19}$$

According to the convention (1.4.19) we get that the sequence $\{x^{(p,1)}\}_{p=0}^{\infty}$ is bounded. Thus there exits a convergent subsequence $\{x^{(p_j,1)}\}_{j=1}^{\infty}$, let's denote it as

$$\lim_{j \to \infty} x^{(p_j,1)} = \hat{x}. \tag{1.4.20}$$

But, from (1.4.14) we get

$$x^{(p_j,2)} = x^{(p_j,1)} + \frac{r^{(p_j,2)}}{\|w^{(2)}\|^2} w^{(2)}, \ \forall \, j > 0. \tag{1.4.21}$$

thus, by taking the limit following j and using (1.4.18), (1.4.20)

$$\lim_{j \to \infty} x^{(p_j,2)} = \hat{x}. \tag{1.4.22}$$

With the same way for $i = 3, \cdots, m$ we obtain

$$\lim_{j \to \infty} x^{(p_j,i)} = \hat{x}, \quad \forall \, i = 1, \cdots, m. \tag{1.4.23}$$

Then, from (1.4.23) we get for any $i = 1, \cdots, m$,

$$\lim_{j \to \infty} \langle x^{(p_j,i-1)}, a_{i+1} \rangle - b_{i+1} = \langle \hat{x}, a_{i+1} \rangle - b_{i+1}, \tag{1.4.24}$$

and from (1.4.13) and (1.4.18),

$$\lim_{j \to \infty} \langle x^{(p_j,i-1)}, a_{i+1} \rangle - b_{i+1} = 0. \tag{1.4.25}$$

Thus, from (1.4.24)-(1.4.25) it results in

$$\langle \hat{x}, a_{i+1} \rangle - b_{i+1} = 0, \quad \forall\, i = 1, \cdots, m, \tag{1.4.26}$$

that is

$$A\hat{x} = b. \tag{1.4.27}$$

With the use of the iterative relations

$$x^{(0,1)} = x^{(0,0)} + \frac{b_1 - \langle a_1, x^{(0,0)} \rangle}{\|w^{(1)}\|^2} w^{(1)},$$

and

$$x^{(p,i)} = x^{(p,i-1)} + \frac{r^{(p,i)}}{\|w^{(i)}\|^2} w^{(i)},$$

$w^{(i)}$ and $w^{(1)}$ are defined in (1.4.8). It is easy to deduce that $P_{\mathcal{N}(A)}(x^{(k)}) = P_{\mathcal{N}(A)}(x^{(0)})$, and so

$$P_{\mathcal{N}(A)}(\hat{x}) = P_{\mathcal{N}(A)}(x^{(0)}). \tag{1.4.28}$$

From the hypothesis of the theorem, we know that

$$A\tilde{x} = b, \quad P_{\mathcal{N}(A)}(\tilde{x}) = P_{\mathcal{N}(A)}(x^{(0)}). \tag{1.4.29}$$

By (1.4.27), (1.4.28) and (1.4.29), we get

$$\lim_{j\to\infty} x^{(p_j,i)} - \hat{x} = \lim_{j\to\infty} x^{(p_j,i)} - \tilde{x} = 0, \quad \forall\, i = 1, 2, \cdots, m.$$

If we set $k_j = p_j \cdot m + i$, then $\lim_{j\to\infty} \|x^{(k_j)} - \tilde{x}\| = \alpha = 0$. Based on monotonicity, $\lim_{k\to\infty} \|x^{(k)} - \tilde{x}\| = 0$, so the sequence $\{x^{(k)}\}$ is convergent to $\tilde{x}$ based on monotonicity.

In addition, if $x^{(0)} \in \mathcal{R}(A^T)$, then $P_{\mathcal{N}(A)}(x^{(0)}) = 0$ and so $\{x^{(k)}\}$ converges to the least-norm solution of (1.1.1), i.e.,

$$\lim_{k\to\infty} x^{(k)} = x^*.$$

$\square$

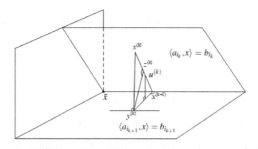

Figure 1.4.2: KO, $m = 3$

Remark 1.4.1. *For Kaczmarz method, it holds that* $\|x^{(k+1)} - \tilde{x}\|^2 = \|x^{(k)} - \tilde{x}\|^2 - \dfrac{|r_{i_{k+1}}^{(k)}|^2}{\|a_{i_{k+1}}\|^2}$, *and the KO holds that* $\|x^{(k+1)} - \tilde{x}\|^2 = \|x^{(k)} - \tilde{x}\|^2 - \dfrac{|r_{i_{k+1}}^{(k)}|^2}{\|a_{i_{k+1}}\|^2} \dfrac{1}{\sin^2(\theta_{i_{k+1}})}$. *So the KO is faster than Kaczmarz method if* $0 < \theta_{i_{k+1}} < \pi/2$.

Remark 1.4.2. *When the coefficient matrix A is a row orthogonal matrix, the KO algorithm degenerates to the Kaczmarz algorithm (right now, $y^{(k)} = z^{(k)} = x^{(k+1)}$). Generally, we use the KO in two ways: one is the online mode, the other is the preprocessing mode.*

(1) Online mode. Each iteration only uses all the information of two adjacent equations, and there is no preprocessing information. Considering that the information of one equation is shared by two adjacent iterations, the KO takes about $10n + 2$ flops per iteration step. In this case, the Kaczmarz method algorithm needs $6n - 1$ floating-point operations per step.

(2) Preprocessing mode. Because of the norm of row vector of matrix A, the inner product of two adjacent row vectors, the direction $w^{(i_k)}$ and its norm are fixed, which can be calculated in advance. After preprocessing(assuming that the above quantity has been calculated), the amount of floating-point number operation of the KO in each step is $4n + 1$ (only $r_{i_{k+1}}^{(k)}$ and $\alpha_{i_k}^{(k)}$ need to be calculated). In this case, the workload of the KO per iterative step is the same as that of Kaczmarz method. But the total cost

of pretreatment for the KO is about $6mn + m$, while that for the Kaczmarz method is $2mn - 1$.

Remark 1.4.3. *Although the workload of the KO in each step is more than or equals that of the Kaczmarz method, compared with the Kaczmarz method, the iteration steps of the KO are significantly reduced, especially for those problems which contain some linear equations with near linear correlation. See Example 1.4.1.*

Example 1.4.1. Consider the following systems of linear equations with two equations

$$\begin{cases} 7x_1 - 8x_2 &= -1, \\ 8x_1 - 7x_2 &= 1 \end{cases} \tag{1.4.30}$$

and

$$\begin{cases} 7x_1 + 8x_2 &= 15, \\ 140x_1 + 159x_2 &= 299. \end{cases} \tag{1.4.31}$$

The two equations in systems (1.4.30) are close to correlation. So if Kaczmarz method is used, 817 steps are needed for the system (1.4.30) and 940627 steps are needed for the system (1.4.31)(two equations are close to correlation) to reach the error requirement $\|x^{(k)} - x^*\| \leqslant \frac{1}{2} \times 10^{-6}$; but with the use of the KO, both systems need only one step to get the exact solution.

Remark 1.4.4. *If we choose three hyper-planes, for example, $\langle a_{i-1}, x \rangle = b_{i-1}$, $\langle a_i, x \rangle = b_i$, and $\langle a_{i+1}, x \rangle = b_{i+1}$, how can we find new method more quickly? If we choose the columns of A, is there a similar convergence method? Are there similar iterative convergence methods for stochastic kaczmarz classes?*

1.5   Coordinate Descent Method with Oblique Direction

Suppose d is a combination of the following two coordinate directions

$$d = e_{i_{k+1}} - \frac{\langle A_{i_{k+1}}, A_{i_k} \rangle}{||A_{i_k}||^2} e_{i_k},$$

where $e_i \in \mathcal{R}^n$, $i = 1, 2, \cdots, n$. Using the system $A^T A x = A^T b$, similar to the equation (1.2.2) and (1.2.4), we get

$$x^{(k+1)} = x^{(k)} + \frac{(A^T b - A^T A x^{(k)})^T (e_{i_{k+1}} - \frac{\langle A_{i_{k+1}}, A_{i_k} \rangle}{||A_{i_k}||^2} e_{i_k})}{\left\| A(e_{i_{k+1}} - \frac{\langle A_{i_{k+1}}, A_{i_k} \rangle}{||A_{i_k}||^2} e_{i_k}) \right\|^2}$$

$$\cdot (e_{i_{k+1}} - \frac{\langle A_{i_{k+1}}, A_{i_k} \rangle}{||A_{i_k}||^2} e_{i_k})$$

$$= x^{(k)} + \frac{A_{i_{k+1}}^T r^{(k)} - \frac{\langle A_{i_{k+1}}, A_{i_k} \rangle}{||A_{i_k}||^2} A_{i_k}^T r^{(k)}}{||A_{i_{k+1}}||^2 - \frac{\langle A_{i_{k+1}}, A_{i_k} \rangle^2}{||A_{i_k}||^2}} (e_{i_{k+1}} - \frac{\langle A_{i_{k+1}}, A_{i_k} \rangle}{||A_{i_k}||^2} e_{i_k}).$$

$$(1.5.1)$$

Now we prove that $A_{i_k}^T r^{(k)} = 0$.

$$A_{i_k}^T r^{(k)} = \langle A_{i_k}, b - A x^{(k)} \rangle$$

$$= \langle A_{i_k}, r^{(k-1)} \rangle - \langle A_{i_k}, r^{(k-1)} \rangle + \frac{\langle A_{i_k}, A_{i_{k-1}} \rangle}{||A_{i_{k-1}}||^2} A_{i_{k-1}}^T r^{(k-1)}$$

$$= \frac{\langle A_{i_k}, A_{i_{k-1}} \rangle}{||A_{i_{k-1}}||^2} A_{i_{k-1}}^T r^{(k-1)}, \quad k = 2, 3, \cdots.$$

We only need to guarantee $A_{i_1}^T r^{(1)} = 0$, so we need to take the simplest coordinate descent projection as the first step. Then the equation (1.5.1) becomes

$$x^{(k+1)} = x^{(k)} + \frac{A_{i_{k+1}}^T r^{(k)}}{||A_{i_{k+1}}||^2 - \frac{\langle A_{i_{k+1}}, A_{i_k} \rangle^2}{||A_{i_k}||^2}} (e_{i_{k+1}} - \frac{\langle A_{i_{k+1}}, A_{i_k} \rangle}{||A_{i_k}||^2} e_{i_k}).$$

Algorithm 1.5.1 Coordinate Descent Method with Oblique Direction (CDO)

---

 procedure $(A,\ b,\ x^{(0)},\ K,\ \varepsilon > 0)$

  For $i = 1 : n$, $N(i) = \|A_i\|^2$

  Compute $r^{(0)} = b - Ax^{(0)}$, $\alpha_0 = \frac{\langle A_1, r^{(0)} \rangle}{N(1)}$, $x^{(1)} = x^{(0)} + \alpha_0 e_1$, $r^{(1)} = r^{(0)} - \alpha_0 A_1$, and set $i_{k+1} = 1$

  for $k = 1, 2, \cdots, K$ **do**

   Set $i_k = i_{k+1}$ and choose a new i_{k+1}: $i_{k+1} = mod(k, n) + 1$

   Compute $G_{i_k} = \langle A_{i_k}, A_{i_{k+1}} \rangle$ and $g_{i_k} = N(i_{k+1}) - \frac{G_{i_k}}{N(i_k)} G_{i_k}$

   if $g_{i_k} > \varepsilon$ **then**

    Compute $\alpha_k = \frac{\langle A_{i_{k+1}}, r^{(k)} \rangle}{g_{i_k}}$ and $\beta_k = -\frac{G(i_k)}{N(i_k)} \alpha_k$

    Compute $x^{(k+1)} = x^{(k)} + \alpha_k e_{i_{k+1}} + \beta_k e_{i_k}$, and $r^{(k+1)} = r^{(k)} - \alpha_k A_{i_{k+1}} - \beta_k A_{i_k}$

   end if

  end for

  Output: $x^{(K+1)}$

 end procedure

The algorithm is described in Algorithm 1.5.1. Without losing generality, we assume that all columns of A are not zero vectors.

It's easy to get

$$
\begin{aligned}
A_{i_{k-1}}^T r^{(k)} &= A_{i_{k-1}}^T (r^{(k-1)} - \alpha_{k-1} A_{i_k} - \beta_{k-1} A_{i_{k-1}}) \\
&= A_{i_{k-1}}^T (r^{(k-1)} - \frac{\langle A_{i_k}, r^{(k-1)} \rangle}{g_{i_{k-1}}} A_{i_k} + \frac{\langle A_{i_{k-1}}, A_{i_k} \rangle \langle A_{i_k}, r^{(k-1)} \rangle}{||A_{i_{k-1}}||^2 g_{i_k}} A_{i_{k-1}}) \\
&= 0. \quad k = 2, 3, \cdots
\end{aligned}
$$

The last equality holds due to $A_{i_k}^T r^{(k)} = 0, k = 1, 2, \cdots$.

Consider a linear least-squares problem

$$
\underset{x \in \mathcal{R}^n}{arg \min} \|Ax - b\|^2, \tag{1.5.2}
$$

where $b \in \mathcal{R}^m$ is a real m dimensional vector. Let the columns of coefficient matrix $A \in \mathcal{R}^{m \times n}$ be non-zero, which doesn't lose the generality of matrix A. Solving (1.5.2) is equivalent to solving the following normal equation

$$
A^T A x = A^T b. \tag{1.5.3}
$$

Similar to the proof of Theorem 1.4.1, we can get the following result.

Theorem 1.5.1. *Suppose $\tilde{x}$ is a least-squares solution of the problem (1.5.2). Let $x^{(0)} \in \mathcal{R}^n$ be an arbitrary initial approximation, then the sequence $\{x^{(k)}\}_{k=1}^{\infty}$ generated by the CDO algorithm is convergent, and satisfies*

$$
\lim_{k \to \infty} \|x^{(k)} - \tilde{x}\|_{A^T A} = 0. \tag{1.5.4}
$$

Proof. Omitted. □

1.6 General Oblique Projection Method

Oblique projections have been used in the past in several contexts. Kayalar and Weiner[21] promoted oblique projections for local processing in sensor arrays and credit. Murray[22] and Lorch[23] proposed pioneering work on oblique projections. Behrens and Scharf[24] used oblique projections for signal processing applications. Oblique projections in Cimmino's algorithm had been proposed and used by Arioli et al.[25], where the projection is oblique with respect to a given symmetric positive definite matrix G which will be explained next. Censor, Gordon, and Gordon[26] modified the Cimmino's algorithm further by replacing the orthogonal projections onto the hyperplanes $H_i = \{x \in \mathcal{R}^n \,|\, \langle a_i, x \rangle = b_i\}$ by certain oblique projections induced by a given symmetric positive definite weight matrix G.

The energy scalar product and ellipsoidal norm were defined by (Bertsekas and Tsitsiklis[27], Popa[28])

$$\langle x, y \rangle_G := \langle Gx, y \rangle, \tag{1.6.1}$$

and

$$\|x\|_G^2 := \langle x, Gx \rangle. \tag{1.6.2}$$

Given a point $z \in \mathcal{R}^n$, the oblique projection of z onto H_i with respect to G is the unique point $P_{H_i}^G(z) \in H_i$ in which

$$P_{H_i}^G(z) := argmin\{\|x - z\|_G \,|\, x \in H_i\}. \tag{1.6.3}$$

Solving this minimization problem leads to

$$P_{H_i}^G(z) = z + \frac{b_i - \langle a_i, z \rangle}{\|a_i\|_{G^{-1}}^2} G^{-1} a_i, \tag{1.6.4}$$

where G^{-1} is the inverse of G. For $G = I$, the identity matrix, (1.6.3) yields the orthogonal projection of z onto H_i, as given in (1.1.4). The Kaczmarz method with oblique projection is expressed as follows:

$$x^{(k+1)} = x^{(k)} + \frac{b_{i_k} - \langle a_{i_k}, x^{(k)} \rangle}{\|a_{i_k}\|_{G^{-1}}^2} G^{-1} a_{i_k}. \tag{1.6.5}$$

Let G be an $n \times n$ symmetric positive semi-definite matrix and define an energy scalar semi-product and a semi-norm as (1.6.1) and (1.6.2). Let G^+ be the Moore-Penrose pseudoinverse of G and it is supposed that

$\langle G^+ a_i, a_i \rangle \neq 0$, then the generalized oblique projection is given as follows (solving minimization problem (1.6.3) with semi-definite matrix G)

$$P_{H_i}^G(z) = z + \frac{b_i - \langle a_i, z \rangle}{\|a_i\|_{G^+}^2} G^+ a_i, \qquad (1.6.6)$$

In[13] , Censor and Elfving formulated a block-iterative algorithmic scheme for the solution of systems of linear inequalities and/or equations and analyzed its convergence which uses certain generalized oblique projections and diagonal weighting matrices which reflect the sparsity of the underlying matrix of the linear system.

In[29], Lorenz, Rose and Schöpfer considered a randomized Kaczmarz method with mismatched adjoint (RKMA), $V \approx A^T$, that is, $v_j \approx A_j \in \mathcal{R}^m$, $j = 1, 2, \cdots, n$, where $V = (v_1, v_2, \cdots, v_m) \in \mathcal{R}^{n \times m}$. The difference between this and the standard randomized Kaczmarz method is that the usual projection step $x^{(k+1)} = x^{(k)} + \frac{b_{i_k} - \langle a_{i_k}, x^{(k)} \rangle}{\|a_{i_k}\|_2^2} a_{i_k}$ is replaced by

$$x^{(k+1)} = x^{(k)} + \frac{b_{i_k} - \langle a_{i_k}, x^{(k)} \rangle}{\langle a_{i_k}, v_{i_k} \rangle} v_{i_k}. \qquad (1.6.7)$$

This results in $\langle a_{i_k}, x^{(k+1)} \rangle = b_{i_k}$, i.e., the next iterate $x^{(k+1)}$ is on the hyperplanes defined by the i_k-th equation of the system. But since v_{i_k} is not orthogonal to this hyperplane, this is an oblique projection, instead of an orthogonal projection as it would be in the original Kaczmarz method (see Fig.1.6.1). Further let $p_{i_k} > 0, i_k = 1, \ldots, m$ be probabilities and denote $D = diag \left(\frac{p_{i_1}}{\langle a_{i_1}, v_{i_1} \rangle}, \cdots, \frac{p_{i_m}}{\langle a_{i_m}, v_{i_m} \rangle} \right)$ and $S = diag \left(\frac{\|v_{i_1}\|_2^2}{\langle a_{i_1}, v_{i_1} \rangle}, \cdots, \frac{\|v_{i_m}\|_2^2}{\langle a_{i_m}, v_{i_m} \rangle} \right)$. If i_k is randomly chosen with probability p_{i_k} then it holds that

$$E(x^{(k+1)} - x) = (I - V^T D A)(x^{(k)} - x)$$

and

$$E(\|x^{(k+1)} - x\|_2^2) = (1 - \lambda)\|x^{(k)} - x\|_2^2,$$

if $\lambda = \lambda_{\min}(V^T D A + A^T D A - A^T S D A) > 0$.

In[30], under the assumption of matrix standardization, Needell and Ward adopted the two-step iterative method and gave an extension of the

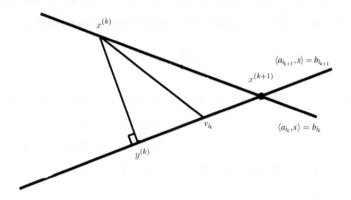

Figure 1.6.1: Oblique Projection, $m = 2$

Kaczmarz method by two randomly selected rows, and showed that this projection algorithm provides exponential convergence to the solution in expectation. The convergence rate significantly improves upon that of the standard randomized Kaczmarz method when the system has correlated rows. That is, the method performs an intermediate projection firstly: $x^{(k)} = x^{(k-1)} + (b_{i_k} - \langle a_{i_k}, x^{(k-1)} \rangle) a_{i_k}$, and then performs the final projection to update the estimation: $x^{(k+1)} = x^{(k)} + (\beta_k - \langle v_{i_k}, x^{(k)} \rangle) v_{i_k}$, where $v_{i_k} = \frac{a_{i_{k+1}} - \mu_k a_{i_k}}{\sqrt{1 - \mu_k^2}}$ (this v_{i_k} is similar to v_{i_k} in (1.6.7)), $\mu_k = \langle a_{i_k}, a_{i_{k+1}} \rangle$ and $\beta_k = \frac{b_{i_{k+1}} - \mu_k b_{i_k}}{\sqrt{1 - \mu_k^2}}$. In fact, $x^{(k)}$ is the orthogonal projection of $x^{(k-1)}$ on the hyperplane $\langle a_{i_k}, x \rangle = b_{i_k}$, $x^{(k+1)}$ is exactly the oblique projection of $x^{(k)}$ along the direction v_{i_k} and $x^{(k+1)}$ is on the intersection of hyperplane $\langle a_{i_k}, x \rangle = b_{i_k}$ and hyperplane $\langle a_{i_{k+1}}, x \rangle = b_{i_{k+1}}$.

In general, the sequences generated by the Kaczmarz method with oblique projection (1.6.4) or with oblique projection (1.6.6) converge to a solution of (1.1.1), but do not necessarily converge to the minimum norm solution (the minimum norm least square solution)(see [28], Chapter 5). Under certain conditions, the sequences generated by the Kaczmarz method (Gauss-Seidel method) with oblique projection will converge to the minimum norm solution (the minimum norm least square solution)[31].

Assume $x^{(k)}$ is on the hyperplane $\langle a_{i_k}, x \rangle = b_{i_k}$, that is, $\langle a_{i_k}, x^{(k)} \rangle = b_{i_k}$. Let

$$B = 2I - \frac{a_{i_k} a_{i_k}^T}{\|a_{i_k}\|_2^2} - \frac{a_{i_{k+1}} a_{i_{k+1}}^T}{\|a_{i_{k+1}}\|_2^2}. \tag{1.6.8}$$

Obviously, $x^T B x > 0$ if $x \neq 0$ and $a_{i_{k+1}} \nparallel a_{i_k}$, i.e., B is a symmetric positive positive matrix if $a_{i_{k+1}} \nparallel a_{i_k}$.

Set $G = B^{-1}$, we get

$$G^{-1} a_{i_{k+1}} = B a_{i_{k+1}} = a_{i_{k+1}} - \frac{\langle a_{i_k}, a_{i_{k+1}} \rangle}{\|a_{i_k}\|_2^2} a_{i_k} \triangleq v_{i_{k+1}},$$

and then

$$\|a_{i_{k+1}}\|_{G^{-1}}^2 = \|a_{i_{k+1}}\|_2^2 - \frac{\langle a_{i_k}, a_{i_{k+1}} \rangle^2}{\|a_{i_k}\|_2^2} \equiv \langle a_{i_{k+1}}, v_{i_{k+1}} \rangle \equiv \|v_{i_{k+1}}\|_2^2.$$

Therefore, the formula (1.6.5) is equivalent to (1.6.7). That is, the oblique projection point of $x^{(k+1)}$ of $x^{(k)}$ along the direction v_{i_k} is on the intersection of hyperplane $\langle a_{i_k}, x \rangle = b_{i_k}$ and hyperplane $\langle a_{i_{k+1}}, x \rangle = b_{i_{k+1}}$ if an appropriate symmetric positive definite matrix G is selected.

In [30], the two-subspace Kaczmarz method performs an intermediate projection: $y^{(k)} = x^{(k)} + (b_{i_k} - \langle a_{i_k}, x^{(k)} \rangle) a_{i_k}$, that is, $y^{(k)}$ is orthogonally projected onto a hyperplane $\langle a_{i_k}, x \rangle = b_{i_k}$ firstly, and then performs the final projection (oblique projection) to get $x^{(k+1)}$ (on the intersection of hyperplane $\langle a_{i_k}, x \rangle = b_{i_k}$ and hyperplane $\langle a_{i_{k+1}}, x \rangle = b_{i_{k+1}}$).

In fact, it is easier to get the iterative point $x^{(k+1)}$ which is on the intersection of hyperplane $\langle a_{i_k}, x \rangle = b_{i_k}$ and hyperplane $\langle a_{i_{k+1}}, x \rangle = b_{i_{k+1}}$. For the sake of simplicity, let A be a normalized matrix by row, i.e., $\|a_i\| = 1$, $i = 1, \cdots, m$. The algorithm is described in Algorithm 1.6.1.

This algorithm takes about $10n$ flops per iteration.

In fact, assume that $y^{(k)}$ is the orthogonal projection of $x^{(k)}$ on hyperplane $\langle a_p, x \rangle = b_p$ and $z^{(k)}$ is the orthogonal projection of $x^{(k)}$ on hyperplane $\langle a_q, x \rangle = b_q$. Now we can find a unique point at the intersection of the hyperplane $\langle a_p, x \rangle = b_p$ and hyperplane $\langle a_q, x \rangle = b_q$ (see Figure 1.6.2.

Algorithm 1.6.1 A Simplified Two-Subspace Kaczmarz (S2SK)Algorithm

---

procedure $(A,\ b,\ x^{(0)},\ K,\ \varepsilon > 0)$

    for $k = 0, 1, \cdots, K - 1$ **do**

        Select $p = k(mod\, m) + 1$, $q = k(mod\ m) + 2$ (Two distinct rows)

        Compute $\mu_k = \langle a_p, a_q \rangle$ (Compute correlation)

        Compute $r_p = b_p - \langle a_p, x^{(k)} \rangle$, $r_q = b_q - \langle a_q, x^{(k)} \rangle$ and $\Delta = 1 - \mu^2$

        if $\Delta > \varepsilon$ **then**

            $\beta_p = \frac{r_p - \mu r_q}{\Delta}$, $\beta_q = \frac{r_q - \mu r_p}{\Delta}$ and $x^{(k+1)} = x^{(k)} + \beta_p a_p + \beta_q a_q$

        end if

    end for

    Output: $x^{(K)}$

end procedure

Let $x^{(k+1)} = x^{(k)} + \beta_p a_p + \beta_q a_q$, and β_p and β_q be undetermined coefficients. From the properties of orthogonal projection, $y^{(k)} = x^{(k)} + r_p a_p$, and $z^{(k)} = x^{(k)} + r_q a_q$, we know that

$$\begin{cases} (x^{(k+1)} - y^{(k)})^T a_p &= 0, \\ (x^{(k+1)} - z^{(k)})^T a_q &= 0. \end{cases} \tag{1.6.9}$$

Then the undetermined coefficients β_p and β_q can be obtained by the following systems

$$\begin{cases} \beta_p + \mu \beta_q &= r_p, \\ \mu \beta_p + \beta_q &= r_q, \end{cases} \tag{1.6.10}$$

that is, $\beta_p = \frac{r_p - \mu r_q}{\Delta}$ and $\beta_q = \frac{r_q - \mu r_p}{\Delta}$, where $r_p = b_p - \langle a_p, x^{(k)} \rangle$, $r_q = b_q - \langle a_q, x^{(k)} \rangle$ and $\Delta = 1 - \mu^2$. This method can be easily extended to more general cases. See[32].

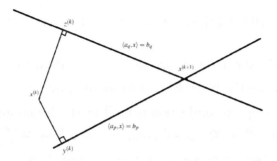

Figure 1.6.2: Simplified two-subspace Kaczmarz, $m = 2$

Let

$$\phi(y) = \frac{1}{2} y^T A A^T y - y^T b. \tag{1.6.11}$$

Select the unit coordinate vectors e_{k_1}, e_{k_2} and compute $y^{(k+1)}$ as

$$y^{(k+1)} = y^{(k)} + \xi e_{k_1} + \eta e_{k_2}, \tag{1.6.12}$$

such that

$$\psi(\xi, \eta) = \phi(y^{(k)} + \xi e_{k_1} + \eta e_{k_2}) \tag{1.6.13}$$

is minimized for some $\xi_k \in \mathcal{R}$ and $\eta_k \in \mathcal{R}$.

By $\frac{\partial \psi}{\partial \xi} = 0$, $\frac{\partial \psi}{\partial \eta} = 0$, $x^{(k)} = A^T y^{(k)}$ and $r^{(k)} = b - Ax^{(k)}$, we know that

$$\begin{cases} \xi_k \|a_{k_1}\|^2 + \eta_k \langle a_{k_1}, a_{k_2} \rangle = r_{k_1}^{(k)}, \\ \xi_k \langle a_{k_2}, a_{k_1} \rangle + \eta_k \|a_{k_2}\|^2 = r_{k_2}^{(k)}. \end{cases} \tag{1.6.14}$$

Therefore,

$$\xi_k = \frac{\|a_{k_2}\|^2 r_{k_1}^{(k)} - \langle a_{k_1}, a_{k_2} \rangle r_{k_2}^{(k)}}{\Delta} \tag{1.6.15}$$

and

$$\eta_k = \frac{\|a_{k_1}\|^2 r_{k_2}^{(k)} - \langle a_{k_1}, a_{k_2} \rangle r_{k_1}^{(k)}}{\Delta}, \tag{1.6.16}$$

if $\Delta = \|a_{k_1}\|^2 \|a_{k_2}\|^2 - \langle a_{k_1}, a_{k_2} \rangle^2 \neq 0$ (Otherwise, a_{k_1} and a_{k_2} are linearly related or coherent). By $x^{(k)} = A^T y^{(k)}$ and equalities (1.6.15), (1.6.16) and

(1.6.12), we get that

$$x^{(k+1)} = x^{(k)} + \xi_k a_{k_1} + \eta_k a_{k_2}. \tag{1.6.17}$$

This is the non-normalized computational form of Algorithm 1.6.1. Furthermore, it is easy to deduce the block Kaczmarz method by this way.

Especially, if one step pre-projection is used at the beginning, that is, when the initial point $x^{(0)} \in \mathcal{R}^n$ is selected, $x^{(1)} = x^{(0)} + \frac{b_1 - \langle a_1, x^{(0)} \rangle}{\|a_1\|^2} a_1$ is computed firstly, and then the cyclic rule can be used to choose a row of the matrix. This method is immediately simplified to a linear search method for solving quadratic optimization problems (1.6.13) with the search direction $p^{(1)} = e_{k_2} + \beta e_{k_1}$. In fact, $r_{k_1}^{(1)} = b_1 - Ax^{(1)} = 0$ by the pre-projection if we choose $e_{k_1} = e_1$ and $e_{k_2} = e_2$. So

$$\psi(\xi, \eta) = \phi(y^{(1)} + \xi e_1 + \eta e_2) = \phi(y^{(1)} + \eta(e_2 + \beta e_1)) = \phi(y^{(1)} + \eta p^{(1)}),$$

where $\beta = \frac{\xi}{\eta} = -\frac{\langle a_1, a_2 \rangle}{\|a_1\|^2}$ by the first equality of (1.6.14) and

$$\eta = \frac{(p^{(1)})^T (b - Ax^{(1)})}{(p^{(1)})^T A p^{(1)}} = \frac{r_2^{(1)}}{\|a_2 + \beta a_1\|^2}.$$

Therefore, by $x = A^T y$, we have

$$x^{(2)} = x^{(1)} + \frac{r_2^{(1)}}{\|a_2 + \beta a_1\|^2} (a_2 + \beta a_1).$$

Next, it can be deduced that $r_2^{(2)} = 0$ if $e_{k_1} = e_2$ and $e_{k_2} = e_3$ are chosen and so

$$x^{(3)} = x^{(2)} + \frac{r_3^{(2)}}{\|a_3 + \beta a_2\|^2} (a_3 + \beta a_2).$$

Similarly, it can be deduced that $r_{k_1}^{(k)} = 0$, if $k_1 = mod(k-1, m) + 1$ and $k_2 = mod(k, m) + 1$ and

$$x^{(k+1)} = x^{(k)} + \frac{r_{k_2}^{(k)}}{\|a_{k_2} + \beta a_{k_1}\|^2} (a_{k_2} + \beta a_{k_1}), \quad k = 1, 2, \cdots.$$

Geometrically, this method is to project the current point onto the intersection of the current plane and the next one. Compared with orthogonal projection, this method is exactly the KO method described in [31] (refer to Figure 1.6.2).

1.7 Block Kaczmarz Method and Block CD Method

At each iteration, the Kaczmarz method makes progress by enforcing a single constraint, while the block Kaczmarz method[7,33] enforces many constraints at once.

Consider the following partitioning of a given matrix A,

$$A = \begin{bmatrix} B_1 \\ B_2 \\ \cdots \\ B_p \end{bmatrix}, \quad 1 \leqslant p \leqslant m. \tag{1.7.1}$$

If $rank(A) = m$, the block Jacobi (BJ) method is described as follows

$$x^{(k+1)} = x^{(k)} + \sum_{i=1}^{p} B_i^T (B_i B_i^T)^{-1} (b_i - B_i x^{(k)}). \tag{1.7.2}$$

The block Kaczmarz (BK) method is obtained by the following iterative step,

$$x^{(k+1)} = x^{(k)} + B_{i_k}^T (B_{i_k} B_{i_k}^T)^{-1} (b_{i_k} - B_{i_k} x^{(k)}), \quad i_k = k \ (mod \ p) + 1. \tag{1.7.3}$$

We now drop the assumption that A has full row rank and assume that the sequence $\{B_i^+\}$ (the gneralized inverse of B_i) can be cheaply computed. In the method, we can also introduce an iteration parameter $\omega_{i_k} \in \mathcal{R}$, as

$$x^{(k+1)} = x^{(k)} + \omega_{i_k} B_{i_k}^+ (b_{i_k} - B_{i_k} x^{(k)}), \quad i_k = k \ (mod \ p) + 1. \tag{1.7.4}$$

If $p = m$ (B_{i_k} is a single row) and $\omega_{i_k} \equiv 1$, this is the Kaczmarz method.

We now describe the methods based on the column-partitioning:

$$A = (A_{\tau_1}, A_{\tau_2}, \cdots, A_{\tau_s}), \quad 1 \leqslant s \leqslant n. \tag{1.7.5}$$

Let $x^T = \left(x_{\tau_1}^T, x_\tau^T, \cdots, x_{\tau_s}^T\right)$ and $I = (I_{\tau_1}, I_{\tau_2}, \cdots, I_{\tau_s})$ be partitions of the vector x and $n \times n$ identity matrix I corresponding to the column index of matrix A. If $rank(A) = n$, the block Coordinate-Descent(BCD) method or block Gauss-Seidel (BGS) method is obtained by following iterative step,

$$x^{(k+1)} = x^{(k)} + I_{\tau_i} \left(A_{\tau_i}^T A_{\tau_i}\right)^{-1} A_{\tau_i}^T r^{(k)}, \quad i = k \ (mod \ s) + 1, \tag{1.7.6}$$

where $r^{(k)} = b - Ax^{(k)}$. Alternatively, the above iteration can also be written as

$$x_{\tau_i}^{(k+1)} = x_{\tau_i}^{(k)} + \left(A_{\tau_i}^T A_{\tau_i}\right)^{-1} A_{\tau_i}^T r^{(k)}, \quad i = k \ (mod \ s) + 1. \tag{1.7.7}$$

We now drop the assumption that A has full column rank and assume that the sequence $\{A_{\tau_i}^+\}$ (the gneralized inverse of A_{τ_i}) can be cheaply computed. Then in the method, we can also introduce an iteration parameter $\omega_i \in \mathcal{R}$, as

$$x_{\tau_i}^{(k+1)} = x_{\tau_i}^{(k)} + \omega_i A_{\tau_i}^+ r^{(k)}, \quad i = k \ (mod \ s) + 1, \tag{1.7.8}$$

where $r^{(k)} = b - Ax^{(k)}$. If $s = n$ (A_{τ_i} is a single column) and $w_i \equiv 1$, this is the Coordinate-Descent (CD) method or Gauss-Seidel (GS) method.

1.8   Greedy Block Kaczmarz Method

The greedy block Kaczmarz (GBK) algorithm constructs control index set and chooses row submatrix in each iteration using a greedy strategy. This algorithm has a linear rate of convergence that can be expressed in terms of the geometric properties of the matrix and its row submatrices. The theoretical analysis and numerical results show that the GBK algorithm can be more efficient than the greedy randomized Kaczmarz (GRK) algorithm (see next chapter) if parameter η is chosen appropriately[34].

Algorithm 1.8.1 Greedy Block Kaczmarz (GBK) Algorithm

---

procedure $(A,\ b,\ x^{(0)},\ K,\ \eta \in (0,1])$

    for $k = 0,1,\cdots,K$ **do**

        Compute $\epsilon_k = \eta \max\limits_{1 \leqslant i \leqslant m} \left\{ \dfrac{|b_i - \langle a_i, x^{(k)} \rangle|^2}{\|a_i\|^2} \right\}$

        Determine the control index set

$$\mathcal{J}_k = \{ i_k : |b_{i_k} - \langle a_{i_k}, x^{(k)} \rangle| \geqslant \epsilon_k \|a_{i_k}\|^2 \} \qquad (1.8.1)$$

        Compute

$$x^{(k+1)} = x^{(k)} + A_{\mathcal{J}_k}^+ \left(b_{\mathcal{J}_k} - A_{\mathcal{J}_k} x^{(k)} \right) \qquad (1.8.2)$$

    end for

    Output: $x^{(K+1)}$

end procedure

Suppose we have approximated the maximum distance (MD) rules[35] in which there is a parameter $\eta \in (0,1]$ for the selection of index i_k such that

$$\frac{|b_{i_k} - \langle a_{i_k}, x^{(k)} \rangle|^2}{\|a_{i_k}\|^2} \geqslant \eta \max_{1 \leqslant i \leqslant m} \left\{ \frac{|b_i - \langle a_i, x^{(k)} \rangle|^2}{\|a_i\|^2} \right\}.$$

Since the index i_k satisfying the above inequality may not be unique, we let $\mathcal{J}_k = \{ i_k : |b_{i_k} - \langle a_{i_k}, x^{(k)} \rangle| \geqslant \epsilon_k \|a_{i_k}\|^2 \}$, where $\epsilon_k = \eta \max\limits_{1 \leqslant i \leqslant m} \left\{ \frac{|b_i - \langle a_i, x^{(k)} \rangle|^2}{\|a_i\|^2} \right\}$. We then select a row index subset $\mathcal{J}_k$ of the matrix A, and project the current iterate point $x^{(k)}$ orthogonally onto the solution space of $\{ x | A_{\mathcal{J}_k} x = b_{\mathcal{J}_k} \}$, where $A_{\mathcal{J}_k}$ and $A_{\mathcal{J}_k}^+$ stand for the row submatrix of A indexed by $\mathcal{J}_k$, i.e. $A_{\mathcal{J}_k} = A(\mathcal{J}_k, :)$ and its Moore-Penrose generalized inverse, respectively; while $b_{\mathcal{J}_k}$ is the subvector of b with components listed in $\mathcal{J}_k$. According to this greedy strategy, we can describe the greedy block Kaczmarz algorithm framework, named Algorithm 1.8.1.

Remark 1.8.1. *The control index set $\mathcal{J}_k \subseteq [m]$ is nonempty. Indeed, if*

$$\frac{|b_j - \langle a_j, x^{(k)} \rangle|^2}{\|a_j\|^2} = \max_{1 \leqslant i \leqslant m} \left\{ \frac{|b_i - \langle a_i, x^{(k)} \rangle|^2}{\|a_i\|^2} \right\},$$

then $j \in \mathcal{J}_k$.

The following theorem describes the convergence theory of the greedy block Kaczmarz algorithm.

Theorem 1.8.1. *Let the linear system $Ax = b$ be consistent. Then the iteration sequence $\{x^{(k)}\}$ generated by Algorithm 1.8.1 converges to the minimal norm solution $x^* = A^+ b$. Moreover we have for any $k \geqslant 0$ that*

$$\|x^{(k+1)} - x^*\|^2 \leqslant \left(1 - \gamma_k(\eta) \frac{\sigma_{\min}^2(A)}{\|A\|_F^2} \right) \|x^{(k)} - x^*\|^2, \tag{1.8.3}$$

where $\gamma_k(\eta) = \eta \frac{\|A\|_F^2}{\|A\|_F^2 - \|A_{\mathcal{J}_{k-1}}\|_F^2} \frac{\|A_{\mathcal{J}_k}\|_F^2}{\sigma_{\max}^2(A_{\mathcal{J}_k})}$ (defining $\gamma_0(\eta) \equiv \eta \frac{\|A_{\mathcal{J}_0}\|_F^2}{\sigma_{\max}^2(A_{\mathcal{J}_0})}$), $\eta \in (0, 1]$, and $\sigma_{\min}(A)$ and $\sigma_{\max}(A)$ are the minimum and maximum singular values of A, respectively.

Proof. Let $e_k = x^{(k)} - x^*$. According to the update rule (1.8.2), we have

$$x^{(k+1)} = x^{(k)} + A_{\mathcal{J}_k}^+ \left(A_{\mathcal{J}_k} x^* - A_{\mathcal{J}_k} x^{(k)} \right).$$

Subtract x^* from the both sides of the above identity, we obtain

$$e_{k+1} = \left(I - A_{\mathcal{J}_k}^+ A_{\mathcal{J}_k} \right) e_k.$$

Since $A_{\mathcal{J}_k}^+ A_{\mathcal{J}_k}$ is an orthogonal projector, using the Pythagorean Theorem we have the following relation

$$\|e_{k+1}\|^2 = \| \left(I - A_{\mathcal{J}_k}^+ A_{\mathcal{J}_k} \right) e_k \|^2 = \|e_k\|^2 - \|A_{\mathcal{J}_k}^+ A_{\mathcal{J}_k} e_k\|^2. \tag{1.8.4}$$

Be aware that

$$\begin{aligned}
\|A_{\mathcal{J}_k}^+ A_{\mathcal{J}_k} e_k\|^2 &\geqslant \sigma_{\min}^2(A_{\mathcal{J}_k}^+) \|A_{\mathcal{J}_k} e_k\|^2 \\
&= \sigma_{\min}^2(A_{\mathcal{J}_k}^+) \sum_{i_k \in \mathcal{J}_k} |\langle a_{i_k}, e_k \rangle|^2 \\
&\geqslant \sigma_{\min}^2(A_{\mathcal{J}_k}^+) \sum_{i_k \in \mathcal{J}_k} \epsilon_k \|a_{i_k}\|^2 \\
&= \frac{\|A_{\mathcal{J}_k}\|_F^2}{\sigma_{\max}^2(A_{\mathcal{J}_k})} \epsilon_k,
\end{aligned}$$

where the second inequality depends on (1.8.1) and the last equality holds because of the fact $\sigma_{\min}^2(A_{\mathcal{J}_k}^+) = \sigma_{\max}^{-2}(A_{\mathcal{J}_k})$. Furthermore, from Algorithm 1.8.1, we have

$$
\begin{aligned}
b - Ax^{(k)} &= b - A\left(x^{(k-1)} + A_{\mathcal{J}_{k-1}}^+(b_{\mathcal{J}_{k-1}} - A_{\mathcal{J}_{k-1}}x^{(k-1)})\right) \\
&= (b - Ax^{(k-1)}) - AA_{\mathcal{J}_{k-1}}^+(b_{\mathcal{J}_{k-1}} - A_{\mathcal{J}_{k-1}}x^{(k-1)}), \ k = 1, 2, \cdots.
\end{aligned}
$$

Thus

$$
b_{\mathcal{J}_{k-1}} - A_{\mathcal{J}_{k-1}}x^{(k)} = (b_{\mathcal{J}_{k-1}} - A_{\mathcal{J}_{k-1}}x^{(k-1)}) - A_{\mathcal{J}_{k-1}}A_{\mathcal{J}_{k-1}}^+(b_{\mathcal{J}_{k-1}} - A_{\mathcal{J}_{k-1}}x^{(k-1)}) = 0.
$$

For $k = 1, 2, \cdots,$, we know that

$$
\begin{aligned}
\|b - Ax^{(k)}\|^2 &= \sum_{i \in [m] \setminus \mathcal{J}_{k-1}} |b_i - A_i x^{(k)}|^2 \\
&= \sum_{i \in [m] \setminus \mathcal{J}_{k-1}} \frac{|b_i - A_i x^{(k)}|^2}{\|A_i\|^2} \|A_i\|^2 \\
&\leqslant \max_{1 \leqslant i \leqslant m} \left\{ \frac{|b_i - A_i x^{(k)}|^2}{\|A_i\|^2} \right\} \left(\|A\|_F^2 - \|A_{\mathcal{J}_{k-1}}\|_F^2 \right).
\end{aligned}
$$

Hence

$$
\begin{aligned}
\epsilon_k &= \eta \max_{1 \leqslant i \leqslant m} \left\{ \frac{|b_i - A_i x^{(k)}|^2}{\|A_i\|^2} \right\} \\
&\geqslant \eta \frac{\|b - Ax^{(k)}\|^2}{\|A\|_F^2 - \|A_{\mathcal{J}_{k-1}}\|_F^2} \\
&\geqslant \eta \frac{\|A\|_F^2}{\|A\|_F^2 - \|A_{\mathcal{J}_{k-1}}\|_F^2} \frac{\sigma_{\min}^2(A)}{\|A\|_F^2} \|e_k\|^2,
\end{aligned}
$$

where the second inequality holds because the $\|Ae_k\|^2 \geqslant \sigma_{\min}^2(A)\|e_k\|^2$. Hence

$$
\|A_{\mathcal{J}_k}^+ A_{\mathcal{J}_k} e_k\|^2 \geqslant \eta \frac{\|A\|_F^2}{\|A\|_F^2 - \|A_{\mathcal{J}_{k-1}}\|_F^2} \frac{\|A_{\mathcal{J}_k}\|_F^2}{\sigma_{\max}^2(A_{\mathcal{J}_k})} \frac{\sigma_{\min}^2(A)}{\|A\|_F^2} \|e_k\|^2. \quad (1.8.5)
$$

Combining (1.8.4) and (1.8.5), we finally have the inequality (1.8.3). $\qquad\square$

Remark 1.8.2. *In Algorithm 1.8.1, when parameter*

$$
\eta = \hat{\eta} \triangleq \frac{1}{2} + \frac{1}{2} \frac{\|b - Ax^{(k)}\|^2}{\|A\|_F^2} \left(\max_{1 \leqslant i \leqslant m} \left\{ \frac{|b_i - A_i x^{(k)}|^2}{\|A_i\|^2} \right\} \right)^{-1},
$$

it is easy to calculate $\epsilon_k = \hat{\epsilon}_k \triangleq \frac{1}{2}\left(\max\limits_{1\leqslant i\leqslant m}\left\{\frac{|b_i - A_i x^{(k)}|^2}{\|A_i\|^2}\right\} + \frac{\|b - Ax^{(k)}\|^2}{\|A\|_F^2}\right)$ *and*

$\hat{J}_k = \{i_k : |b_{i_k} - A_{i_k}x^{(k)}|^2 \geqslant \hat{\epsilon}_k \|A_{i_k}\|^2\}$. *In this case, we have the following error estimate*

$$\|x^{(k+1)} - x^*\|^2 \leqslant \left(1 - \gamma_k \frac{\|A_{\hat{J}_k}\|_F^2}{\sigma_{\max}^2(A_{\hat{J}_k})} \frac{\sigma_{\min}^2(A)}{\|A\|_F^2}\right)\|x^{(k)} - x^*\|^2 \qquad (1.8.6)$$

where $\gamma_k = \frac{1}{2}\left(\frac{\|A\|_F^2}{\|A\|_F^2 - \|A_{\hat{J}_{k-1}}\|_F^2} + 1\right)$, *defining* $\gamma_0 \equiv 1$. *Note that the greedy randomized Kaczmarz (GRK) algorithm in [36] for solving the linear systems (1.1.1) is based on the control index set* $\hat{J}_k$ *and its error estimate in expectation reads*

$$\|x^{(k+1)} - x^*\|^2 \leqslant \left(1 - \gamma \frac{\sigma_{\min}^2(A)}{\|A\|_F^2}\right)\|x^{(k)} - x^*\|^2 \qquad (1.8.7)$$

where $\gamma_k = \frac{1}{2}\left(\frac{\|A\|_F^2}{\|A\|_F^2 - \min\limits_{1\leqslant i\leqslant m}\|A_i\|^2} + 1\right)$. *The comparison of both convergent factors in these two algorithms will be given in Remark 1.8.3.*

The error estimate (1.8.3) for convergence of Algorithm 1.8.1 exhibits the linear rate of convergence of the GBK method. Obviously, the geometric properties of the matrix and its row submatrices control the rate of convergence. The GBK method can be used to solve any compatible linear system regardless of whether it is ill-conditioned or not, but the bigger the smallest singular value $\sigma_{\min}(A)$ of A is, the faster its convergence rate is.

Remark 1.8.3. *We further remark that the factor of convergence of the GBK algorithm is smaller, uniformly with respect to the iteration step* k, *than that of the GRK algorithm in [36], which shows theoretically that the GBK algorithm is superior to the GRK on their convergence rate though the GRK method has many other advantages.*

In fact, as $\|A_{\hat{J}_k}\|_F^2 \geqslant \sigma_{\max}^2(A_{\hat{J}_k})$ *and* $\min\limits_{1\leqslant j\leqslant m}\|A_j\|^2 \leqslant \|A_{\hat{J}_{k-1}}\|_F^2$, *it holds that*

$$\frac{\|A\|_F^2}{\|A\|_F^2 - \min\limits_{1\leqslant j\leqslant m}\|A_j\|^2} \leqslant \frac{\|A\|_F^2}{\|A\|_F^2 - \|A_{\hat{J}_{k-1}}\|_F^2},$$

i.e., $\gamma_k \geqslant \gamma$. In addition, we know that the inequality $\dfrac{\|A_{\hat{\mathcal{J}}_k}\|_F^2}{\sigma_{\max}^2(A_{\hat{\mathcal{J}}_k})} \geqslant 1$ always holds. Hence,

$$1 - \gamma_k \frac{\|A_{\hat{\mathcal{J}}_k}\|_F^2}{\sigma_{\max}^2(A_{\hat{\mathcal{J}}_k})} \frac{\sigma_{\min}^2(A)}{\|A\|_F^2} \leqslant 1 - \gamma \frac{\sigma_{\min}^2(A)}{\|A\|_F^2}.$$

1.9   Greedy Block Gauss-Seidel Method

Similar to GBK method,[37] presented a greedy block Gauss-Seidel (G-BGS) method (or greedy block coordinate descent (GBCD) method) for solving large linear least squares problem. Theoretical analysis demonstrates that the convergence factor of the GBGS method can be much smaller than that of the greedy randomized coordinate descent (GRCD) method.

Algorithm 1.9.1 Greedy Block Gauss-Seidel (GBGS) Algorithm

---

procedure $(A,\ b,\ x^{(0)},\ K)$

  for $k = 1, 2, \cdots, K$ **do**

    Compute $\epsilon_k = \dfrac{\theta}{\|A^T r^{(k)}\|_2^2} \max\limits_{1 \leqslant j \leqslant n} \left\{ \dfrac{|A_j^T r^{(k)}|}{\|A_j\|^2} \right\} + \dfrac{1-\theta}{\|A\|_F^2}$

    Determine the index set of positive integers

$$\mathcal{J}_k = \left\{ j \,\big|\, |A_j^T r^{(k)}|^2 \geqslant \epsilon_k \|A^T r^{(k)}\|^2 \|A_j\|^2 \right\}$$

    Set

$$x^{(k+1)} = x^{(k)} + I_{\mathcal{J}_k} A_{\mathcal{J}_k}^+ r^{(k)} \tag{1.9.1}$$

  end for

  Output: $x^{(K+1)}$

end procedure

There are two main steps in the GBGS method. The first step is to devise a greedy rule to decide the index set $\mathcal{J}_k$ whose specific definition is given in Algorithm 1.9.1, and the second step is to update $x^{(k+1)}$ using a update formula (1.9.1), with which the GBGS method can minimize the objective function through all coordinates from the index set $\mathcal{J}_k$ at the same time. Note that the GRCD method only updates one coordinate in each iteration.

Remark 1.9.1. *Note that if* $\frac{|A_{j_k}^T r^{(k)}|^2}{\|A_{j_k}\|^2} = \max\limits_{1 \leqslant j \leqslant n} \left\{ \frac{|A_j^T r^{(k)}|^2}{\|A_j\|^2} \right\}$, *then* $j_k \in \mathcal{J}_k$. *So the index set $\mathcal{J}_k$ in Algorithm 1.9.1 is nonempty for all iteration index k.*

Theorem 1.9.1. *The iteration sequence $\{x^{(k)}\}$ generated by Algorithm 1.9.1, starting from an initial guess $x^{(0)} \in \mathcal{R}^n$, converges linearly to the unique least squares solution $x^* = A^+ b$ and*

$$\|x^{(1)} - x^*\|_{A^T A}^2 \leqslant \left(1 - \frac{\|A_{\mathcal{J}_0}\|_F^2}{\sigma_{\max}^2(A_{\mathcal{J}_0})} \frac{\sigma_{\min}^2(A)}{\|A\|_F^2} \right) \|x^{(0)} - x^*\|_{A^T A}^2, \qquad (1.9.2)$$

and

$$\|x^{(k+1)} - x^*\|_{A^T A}^2$$
$$\leqslant \left(1 - \frac{\|A_{\mathcal{J}_k}\|_F^2}{\sigma_{\max}^2(A_{\mathcal{J}_k})} \left(\theta \frac{\|A\|_F^2}{\|A\|_F^2 - \|A_{\mathcal{J}_{k-1}}\|_F^2} + (1-\theta) \right) \frac{\sigma_{\min}^2(A)}{\|A\|_F^2} \right) \|x^{(k)} - x^*\|_{A^T A}^2,$$
$$k = 1, 2, \cdots$$

$$(1.9.3)$$

Moreover, let $\alpha = \max\{\sigma_{\max}^2(A_{\mathcal{J}_k})\}$, $\beta = \min\{\|A_{\mathcal{J}_k}\|_F^2\}$, and $\gamma = \max\{\|A\|_F^2 - \|A_{\mathcal{J}_{k-1}}\|_F^2\}$. Then

$$\|x^{(k+1)} - x^*\|_{A^T A}^2 \leqslant \left(1 - \frac{\beta}{\alpha} \left(\theta \frac{\|A\|_F^2}{\gamma} + (1-\theta) \right) \frac{\sigma_{\min}^2(A)}{\|A\|_F^2} \right)^{k-1}$$
$$\times \left(1 - \frac{\|A_{\mathcal{J}_0}\|_F^2}{\sigma_{\max}^2(A_{\mathcal{J}_0})} \frac{\sigma_{\min}^2(A)}{\|A\|_F^2} \right) \|x^{(0)} - x^*\|_{A^T A}^2, \quad k = 1, 2, \cdots$$
$$(1.9.4)$$

Proof. From the update formula (1.9.1) in Algorithm 1.9.1, we have

$$A(x^{(k+1)} - x^*) = A(x^{(k)} - x^*) + A_{\mathcal{J}_k} A_{\mathcal{J}_k}^+ r^{(k)},$$

which together with the fact $A_{\mathcal{J}_k} A_{\mathcal{J}_k}^+ = (A_{\mathcal{J}_k} A_{\mathcal{J}_k}^+)^T$ and $A^T A x^* = A^T b$ gives

$$A(x^{(k+1)} - x^*) = A(x^{(k)} - x^*) + (A_{\mathcal{J}_k}^+)^T A_{\mathcal{J}_k}^T (b - A x^{(k)})$$
$$= (I - A_{\mathcal{J}_k} A_{\mathcal{J}_k}^+) A(x^{(k)} - x^*).$$

Since $A_{\mathcal{J}_k} A_{\mathcal{J}_k}^+$ is an orthogonal projector, taking the square of the Euclidean norm on both sides of the above equation and applying Pythagorean theorem, we get

$$
\begin{aligned}
\|A(x^{(k+1)} - x^*)\|^2 &= \|(I - A_{\mathcal{J}_k} A_{\mathcal{J}_k}^+) A(x^{(k)} - x^*)\|^2 \\
&= \|A(x^{(k)} - x^*)\|^2 - \|A_{\mathcal{J}_k} A_{\mathcal{J}_k}^+ A(x^{(k)} - x^*)\|^2 \\
&= \|A(x^{(k)} - x^*)\|^2 - \|(A_{\mathcal{J}_k}^+)^T A_{\mathcal{J}_k}^T A(x^{(k)} - x^*)\|^2
\end{aligned}
$$

or equivalently,

$$
\|x^{(k+1)} - x^*\|_{A^T A}^2 = \|x^{(k)} - x^*\|_{A^T A}^2 - \|(A_{\mathcal{J}_k}^+)^T A_{\mathcal{J}_k}^T A(x^{(k)} - x^*)\|^2,
$$

which together with $\|A^T x\|^2 \geqslant \sigma_{\min}^2(A)\|x\|^2$ and $\sigma_{\min}^2(A_{\mathcal{J}_k}^+) = \sigma_{max}^{-2}(A_{\mathcal{J}_k})$ yields

$$
\begin{aligned}
\|x^{(k+1)} - x^*\|_{A^T A}^2 &\leqslant \|x^{(k)} - x^*\|_{A^T A}^2 - \sigma_{min}^2(A_{\mathcal{J}_k}^+)\|A_{\mathcal{J}_k}^T A(x^{(k)} - x^*)\|^2 \\
&= \|x^{(k)} - x^*\|_{A^T A}^2 - \frac{1}{\sigma_{\max}^2(A_{\mathcal{J}_k})}\|A_{\mathcal{J}_k}^T A(x^{(k)} - x^*)\|^2.
\end{aligned}
$$

On the other hand, from Algorithm 1.9.1, we have

$$
\begin{aligned}
\|A_{\mathcal{J}_k}^T A(x^{(k)} - x^*)\|^2 &= \sum_{j \in \mathcal{J}_k} |A_j^T r^{(k)}|^2 \\
&\geqslant \sum_{j \in \mathcal{J}_k} \epsilon_k \|A^T r^{(k)}\|^2 \|A_j\|^2 = \epsilon_k \|A^T r^{(k)}\|^2 \|A_{\mathcal{J}_k}\|_F^2.
\end{aligned}
$$

Then,

$$
\|x^{(k+1)} - x^*\|_{A^T A}^2 \leqslant \|x^{(k)} - x^*\|_{A^T A}^2 - \frac{\|A_{\mathcal{J}_k}\|_F^2}{\sigma_{\max}^2(A_{\mathcal{J}_k})}\|A^T r^{(k)}\|^2 \epsilon_k. \quad (1.9.5)
$$

For $k = 0$, we have

$$
\begin{aligned}
\|A^T r^{(0)}\|^2 \epsilon_0 &= \theta \max_{1 \leqslant j \leqslant n} \left\{ \frac{|A_j^T r^{(0)}|}{\|A_j\|^2} \right\} + (1 - \theta) \frac{\|A^T r^{(0)}\|^2}{\|A\|_F^2} \\
&\geqslant \theta \sum_{j=1}^{n} \frac{\|A_j\|^2}{\|A\|^2} \frac{|A_j^T r^{(0)}|}{\|A_j\|^2} + (1 - \theta) \frac{\|A^T r^{(0)}\|^2}{\|A\|_F^2} \\
&= \theta \frac{\|A^T r^{(0)}\|}{\|A\|^2} + (1 - \theta) \frac{\|A^T r^{(0)}\|^2}{\|A\|_F^2} = \frac{\|A^T r^{(0)}\|^2}{\|A\|_F^2}.
\end{aligned}
$$

which together with $\|A^T x\|^2 \geqslant \sigma_{\min}^2(A)\|x\|^2$ leads to

$$
\|A^T r^{(0)}\|^2 \epsilon_0 \geqslant \frac{\sigma_{\min}^2(A)}{\|A\|_F^2} \|A(x^{(0)} - x^*)\|^2 = \frac{\sigma_{\min}^2(A)}{\|A\|_F^2} \|x^{(0)} - x^*\|_{A^T A}^2. \quad (1.9.6)
$$

Thus, substituting (1.9.6) into (1.9.5), we obtain

$$\|x^{(1)} - x^*\|_{A^T A}^2 \leqslant \|x^{(0)} - x^*\|_{A^T A}^2 - \frac{\|A_{\mathcal{J}_0}\|_F^2}{\sigma_{\max}^2(A_{\mathcal{J}_0})} \frac{\sigma_{\min}^2(A)}{\|A\|_F^2} \|x^{(0)} - x^*\|_{A^T A}^2$$

$$= \left(1 - \frac{\|A_{\mathcal{J}_0}\|_F^2}{\sigma_{\max}^2(A_{\mathcal{J}_0})} \frac{\sigma_{\min}^2(A)}{\|A\|_F^2}\right) \|x^{(0)} - x^*\|_{A^T A}^2,$$

which is just the estimate (1.9.2).

For $k \geqslant 1$, to find the lower bound of $\|A^T r^{(k)}\|^2 \epsilon_k$, we first note that

$$A_{\mathcal{J}_{k-1}}^T r^{(k)} = A_{\mathcal{J}_{k-1}}^T \left(b - A(x^{(k-1)} + I_{\mathcal{J}_{k-1}} A_{\mathcal{J}_{k-1}}^+ r^{(k-1)})\right)$$

$$= A_{\mathcal{J}_{k-1}}^T r^{(k-1)} - A_{\mathcal{J}_{k-1}}^T A_{\mathcal{J}_{k-1}} A_{\mathcal{J}_{k-1}}^+ r^{(k-1)}$$

$$= 0,$$

where we have used the update formula (1.9.1) and the property of the Moore-Penrose pseudoinverse. Then

$$\|A^T r^{(k)}\|^2 = \sum_{j=1}^{n} |A_j^T r^{(k)}|^2 = \sum_{\substack{j=1 \\ j \notin \mathcal{J}_{k-1}}}^{n} |A_j^T r^{(k)}|^2 = \sum_{\substack{j=1 \\ j \notin \mathcal{J}_{k-1}}}^{n} \frac{|A_j^T r^{(k)}|^2}{\|A_j\|^2} \|A_j\|^2$$

$$\leqslant \max_{1 \leqslant j \leqslant n} \frac{|A_j^T r^{(k)}|^2}{\|A_j\|^2} \sum_{\substack{j=1 \\ j \notin \mathcal{J}_{k-1}}}^{n} \|A_j\|^2$$

$$= \max_{1 \leqslant j \leqslant n} \frac{|A_j^T r^{(k)}|^2}{\|A_j\|^2} \left(\|A\|_F^2 - \|A_{\mathcal{J}_{k-1}}\|_F^2\right),$$

which can first imply a lower bound of $\max_{1 \leqslant j \leqslant n} \frac{|A_j^T r^{(k)}|^2}{\|A_j\|^2}$. Then a lower bound of $\|A^T r^{(k)}\|^2 \epsilon_k$ is obtained as follows

$$\|A^T r^{(k)}\|^2 \epsilon_k \geqslant \theta \frac{\|A^T r^{(k)}\|^2}{\|A\|_F^2 - \|A_{\mathcal{J}_{k-1}}\|_F^2} + (1-\theta) \frac{\|A^T r^{(k)}\|^2}{\|A\|_F^2}.$$

Further, considering $\|A^T x\|^2 \geqslant \sigma_{\min}^2(A) \|x\|^2$, we have

$$\|A^T r^{(k)}\|^2 \epsilon_k \geqslant \left(\theta \frac{1}{\|A\|_F^2 - \|A_{\mathcal{J}_{k-1}}\|_F^2} + (1-\theta) \frac{1}{\|A\|_F^2}\right) \sigma_{\min}^2(A) \|A(x^{(k)} - x^*)\|^2$$

$$= \left(\theta \frac{\|A\|_F^2}{\|A\|_F^2 - \|A_{\mathcal{J}_{k-1}}\|_F^2} + (1-\theta)\right) \frac{\sigma_{\min}^2(A)}{\|A\|_F^2} \|x^{(k)} - x^*\|_{A^T A}^2$$

$$\tag{1.9.7}$$

Thus, substituting (1.9.7) into (1.9.5) gives the estimate (1.9.3). By induction on the iteration index k, we can obtain the estimate (1.9.4). $\square$

Remark 1.9.2. *If* $\theta = \frac{1}{2}$, *i.e., the index sets of the GBGS and GRCD methods are the same, the first term in the right side of (1.9.3) reduces to*

$$\eta = 1 - \frac{\|A_{\mathcal{J}_k}\|_F^2}{\sigma_{\max}^2(A_{\mathcal{J}_k})} \frac{1}{2} \left(\frac{\|A\|_F^2}{\|A\|_F^2 - \|A_{\mathcal{J}_{k-1}}\|_F^2} + 1 \right) \frac{\sigma_{\min}^2(A)}{\|A\|_F^2}.$$

Since

$$\|A\|_F^2 - \|A_{\mathcal{J}_{k-1}}\|_F^2 \leqslant \|A\|_F^2 - \min_{1 \leqslant j \leqslant n} \|A_j\|_F^2 \quad and \quad \frac{\|A_{\mathcal{J}_k}\|_F^2}{\sigma_{\max}^2(A_{\mathcal{J}_k})} \geqslant 1,$$

we have

$$\eta \leqslant 1 - \frac{1}{2} \left(\frac{\|A\|_F^2}{\|A\|_F^2 - \min\limits_{1 \leqslant j \leqslant n} \|A_j\|_F^2} + 1 \right) \frac{\sigma_{\min}^2(A)}{\|A\|_F^2}.$$

Note that the error estimate in expectation of the GRCD method given in [38] is

$$E_k \|x^{(k+1)} - x^*\|_{A^T A}^2$$
$$\leqslant \left(1 - \frac{1}{2} \left(\frac{\|A\|_F^2}{\|A\|_F^2 - \min\limits_{1 \leqslant j \leqslant n} \|A_j\|_F^2} + 1 \right) \frac{\sigma_{\min}^2(A)}{\|A\|_F^2} \right) \|x^{(k)} - x^*\|_{A^T A}^2,$$

where $k = 1, 2, \cdots$. *So the convergence factor of the GBGS method is smaller than that of the GRCD method in this case, and the former can be much smaller than the latter because* $\|A\|_F^2 - \|A_{\mathcal{J}_{k-1}}\|_F^2$ *can be much smaller than* $\|A\|_F^2 - \min\limits_{1 \leqslant j \leqslant n} \|A_j\|_F^2$ *and* $\|A_{\mathcal{J}_k}\|_F^2$ *can be much larger than* $\sigma_{\max}^2(A_{\mathcal{J}_k})$.

Remark 1.9.3. *The greedy block Kaczmarz algorithm[34] for solving large-scale linear systems is a deterministic version of the famous greedy randomized Kaczmarz method given in [36] and the greedy block Gauss-Seidel algorithm[37] is a deterministic version of the greedy randomized Gauss-Seidel method[38].*

Chapter 2

Random Kaczmarz Method and Its Variants

2.1 Random Kaczmarz Method

The Random Kaczmarz (RK) is a randomized version of the Kaczmarz method for consistent, overdetermined linear systems. Thomas Strohmer and Roman Vershynin has proved that it converges with expected exponential rate in 2009[39]. Furthermore, its convergence rate does not depend on the number of equations in the system and the solver does not even need to know the whole system, but only a small random part of it.

It still computes

$$x^{(k+1)} = x^{(k)} + \frac{r_i}{\|a_i\|^2} a_i. \tag{2.1.1}$$

for $k = 1, 2, \cdots$, while the i is chosen from $[m]$ at random with probability proportional to $\|a_i\|^2$.

Denote the probability

$$p_i = \frac{\|a_i\|^2}{\|A\|_F^2}, \tag{2.1.2}$$

where $\|A\|_F^2 = \sum_{i=1}^{m} \|a_i\|^2$, and p_i can be calculated before iteration.

Algorithm 2.1.1 Randomized Kaczmarz (RK) Algorithm

---

procedure $(A,\ b,\ x^{(0)},\ K)$

  for $k = 0, 1, \cdots, K$ **do**

    Select $i_k \in [m]$ with probability $\Pr(\text{row}{=}i_k){=}\dfrac{\|a_{i_k}\|^2}{\|A\|_F^2}$

    Set $x^{(k+1)} = x^{(k)} + \dfrac{b_{i_k} - \langle a_{i_k}, x^{(k)} \rangle}{\|a_{i_k}\|^2} a_{i_k}$

  end for

  Output: $x^{(K+1)}$

end procedure

The convergence analysis of the RK algorithm was firstly given by Strohmer and Vershynin[39], and then improved by Zouzias and Freris[40]. The following theorem is a restatement of the main result of[39] without imposing the full column rank assumption.

Theorem 2.1.1. *([40]) Assume that $A \in \mathcal{R}^{m \times n}$, $b \in \mathcal{R}^n$ and $Ax = b$ has a solution. Denote the least-norm solution $x^* := A^+ b$. In exact arithmetic, Algorithm 2.1.1 converges to x^* in mean square:*

$$E\|x^{(k)} - x^*\|^2 \leqslant \left(1 - \frac{1}{\kappa_F^2(A)}\right)^k \|x^{(0)} - x^*\|^2, \quad \forall k > 0. \tag{2.1.3}$$

Proof. By hypothesis, $x^{(k)} \in \mathcal{R}(A^T)$ for any k when $x^{(0)} \in \mathcal{R}(A^T)$; in addition, the same is true for x^* by the definition of M-P pseudoinverse[19]. If $z \in \mathcal{R}(A^T)$, then it holds (see Proposition A in this section)

$$\sum_{j=1}^{m} |\langle z, a_j \rangle|^2 = \|Az\|^2 \geqslant \sigma_{\min}^2(A)\|z\|^2 = \frac{\|z\|^2}{\|A^+\|^2}. \tag{2.1.4}$$

Using the fact that $\|A\|_F^2 = \sum\limits_{j=1}^{m} \|a_j\|^2$, we can write (2.1.4) as

$$\sum_{j=1}^{m} \frac{\|a_j\|^2}{\|A\|_F^2} |\langle z, \frac{a_j}{\|a_j\|} \rangle|^2 \geqslant \frac{\|z\|^2}{\kappa_F^2(A)}, \quad \textit{for all } z \in \mathcal{R}(A^T). \tag{2.1.5}$$

The main point of the proof is to view the left hand side in (2.1.5) as an expectation of some random variable. Namely, we recall that the solution space of the j-th equation of (1.1.1) is the hyperplane $\{y : \langle y, a_j \rangle = b_j\}$, whose normal is $\frac{a_j}{\|a_j\|}$. Define a random vector Z whose values are the normals to all the equations of (1.1.1), with probabilities as in our algorithm:

$$Z = \frac{a_j}{\|a_j\|} \ with \ probability \ \frac{\|a_j\|^2}{\|A\|_F^2}, \ j = 1, 2, \cdots, m. \qquad (2.1.6)$$

Then the inequality (2.1.5) says that

$$E|\langle z, Z \rangle|^2 \geqslant \kappa_F^{-2}(A)\|z\|^2, \ for \ all \ z \in \mathcal{R}(A^T). \qquad (2.1.7)$$

The orthogonal projection P onto the solution space of a random equation of (1.1.1) is given by $Pz = z - \langle z - x^*, Z \rangle Z$.

Now we are ready to analyze our algorithm. We want to show that the error $\|x^{(k)} - x^*\|_2^2$ reduces at each step in average (conditioned on the previous steps) by at least the factor of $(1 - \kappa_F^{-2}(A))$. The next approximation $x^{(k)}$ is computed from $x^{(k-1)}$ as $x^{(k)} = P_k x^{(k-1)}$, where $P_1, P_2, \cdots$ are independent realizations of the random projection P. The vector $x^{(k-1)} - x^{(k)}$ is in the kernel of P_k. It is orthogonal to the solution space of the equation onto which P_k projects, which contains the vector $x^{(k)} - x^*$ (x^* is the solution of all equations). The orthogonality of these two vectors then yields

$$\|x^{(k)} - x^*\|^2 = \|x^{(k-1)} - x^*\|^2 - \|x^{(k-1)} - x^{(k)}\|^2$$

To complete the proof, we have to bound $\|x^{(k-1)} - x^{(k)}\|^2$ from below. By the definition of $x^{(k)}$, we have

$$\|x^{(k-1)} - x^{(k)}\| = \langle x^{(k-1)} - x^*, Z_k \rangle$$

where $Z_1, Z_2, \cdots$ are independent realizations of the random vector Z. Thus

$$\|x^{(k)} - x^*\|^2 \leqslant \left(1 - \left|\langle \frac{x^{(k-1)} - x^*}{\|x^{(k-1)} - x^*\|}, Z_k \rangle\right|^2\right) \|x^{(k-1)} - x^*\|^2.$$

Now we take the expectation of both sides conditional upon the choice of the random vectors $Z_1, Z_2, \cdots, Z_{k-1}$ (hence we fix the choice of the random projections $P_1, P_2, \cdots, P_{k-1}$ and thus the random vectors $x^{(1)}, x^{(2)}, \cdots, x^{(k-1)}$, and we average over the random vector Z_k). Then

$$E_{\{Z_1, \cdots, Z_{k-1}\}} \|x^{(k)} - x^*\|^2$$
$$\leqslant \left(1 - E_{\{Z_1, \cdots, Z_{k-1}\}} \left| \langle \frac{x^{(k-1)} - x^*}{\|x^{(k-1)} - x^*\|}, Z_k \rangle \right|^2 \right) \|x^{(k-1)} - x^*\|^2.$$

By (2.1.7) and the independence,

$$E_{\{Z_1, \cdots, Z_{k-1}\}} \|x^{(k)} - x^*\|^2 \leqslant \left(1 - \kappa_F^{-2}(A)\right) \|x^{(k-1)} - x^*\|^2.$$

Taking the full expectation of both sides, we conclude that

$$E\|x^{(k)} - x^*\|^2 \leqslant \left(1 - \kappa_F^{-2}(A)\right) E\|x^{(k-1)} - x^*\|^2.$$

$\square$

Remark 2.1.1. *In order to simplify the calculation, normalization is often used in many algorithms, while in this case if a_i is normalized, i.e., $\|a_i\|^2 \equiv 1, i = 1, 2, \cdots, m$, then $p_i \equiv \frac{1}{m}$ and the RK loses its advantage(see [41] for the evaluation of this method).*

Remark 2.1.2. *In fact, the proof of convergence does not need the full rank of matrix A, but only needs that the system of equation is consistent (solvable)[40] and $x^{(0)} \in \mathcal{R}(A^T)$.*

For this improvement, the following proposition is used.

Proposition 2.1.1. *Assume that $A \in \mathcal{R}^{m \times n}$. If $z \in \mathcal{R}^n$ and in the row space of A, i. e., $z \in \mathcal{R}(A^T)$, it follows that*

$$\|Az\| \geqslant \sigma_{\min}(A)\|z\|, \tag{2.1.8}$$

where $\sigma_{\min}(A)$ is the smallest nonzero singular value of A.

Proof. Due to $z \in \mathcal{R}(A^T)$, there exists a $w \in \mathcal{R}^m$ such that $z = A^T w$, therefore,

$$\|Az\|^2 = \|AA^T w\|^2 = w^T AA^T AA^T w. \tag{2.1.9}$$

Denote the SVD of A as

$$A = U\Sigma V^T,$$

where $\Sigma = diag(\sigma_1, \cdots, \sigma_r, 0, \cdots, 0) \in \mathcal{R}^{m \times n}$, $U \in \mathcal{R}^{m \times m}$ and $V \in \mathcal{R}^{n \times n}$ are orthogonal matrices. Then

$$w^T AA^T AA^T w = w^T U\Sigma V^T V\Sigma^T U^T U\Sigma V^T V\Sigma^T U^T w = w^T U(\Sigma\Sigma^T)^2 U^T w. \tag{2.1.10}$$

Denote $g = U^T w$ and from (2.1.10), it follows that

$$\|Az\|^2 = w^T AA^T AA^T w = g^T(\Sigma\Sigma^T)^2 g = \sum_{i=1}^{r} \sigma_i^4 g_i^2. \tag{2.1.11}$$

Again from

$$\sigma_{min}^2(A)\|z\|_2^2 = \sigma_{min}^2(A)w^T AA^T w = \sigma_{min}^2(A)w^T U(\Sigma\Sigma^T)U^T w$$

$$= \sigma_{min}^2(A) \sum_{i=1}^{r} \sigma_i^2 g_i^2 \leqslant \sum_{i=1}^{r} \sigma_i^4 g_i^2, \tag{2.1.12}$$

the inequality in the proposition is established immediately. $\qquad\square$

Remark 2.1.3. *If $z \notin \mathcal{R}(A^T)$, the proposition is not true. For example, if*

$$A = \begin{pmatrix} 1 & 0 \\ 0 & 0 \end{pmatrix} \quad and \quad z = \begin{pmatrix} 1 \\ 1 \end{pmatrix},$$

obviously, $\|Az\| = 1$, $\sigma_{min}(A) = 1$ and $\|z\| = \sqrt{2}$, but $\|Az\| < \sigma_{min}(A)\|z\|$.

Remark 2.1.4. *However, the inequality (2.1.8) holds if $rank(A) = n$ (the full column rank assumption on A) even if $z \notin \mathcal{R}(A^T)$. In fact,*

$$\|Az\|^2 = z^T A^T Az = \sum_{i=1}^{r} \sigma_i^2(A)(V^T z)_i^2 = \sum_{i=1}^{n} \sigma_i^2(A)(V^T z)_i^2$$

$$\geqslant \sigma_{min}^2(A) \sum_{i=1}^{n} (V^T z)_i^2 = \sigma_{min}^2(A)\|V^T z\|^2 = \sigma_{min}^2(A)\|z\|^2.$$

2.2   New Random Kaczmarz Method for $p_i = \frac{|r_i|}{\|r\|_1}$

In a sense, the RK method is superior to other Kaczmarz methods, but its random selection strategy is not necessarily optimal. For instance, if $\|a_i\|_2 (i = 1, 2, \cdots, m)$ is a constant $(\|a_1\|_2 = \|a_2\|_2 = \cdots = \|a_m\|_2)$, the RK method selects one row of matrix A with equal probability, losing its advantage. And T. Strohmer and R. Vershynin claimed that their rule is not optimal, and actually they never claimed it to be optimal, see [42].

The New Random Kaczmarz (NewRK)[43] is a new random selection method that randomly selects a row of matrix A according to the distance from the current iteration point to the hyperplane. The current iteration point is preferentially projected to a farther hyperplane, it then avoids the problems mentioned above and achieves the purpose of accelerating convergence.

The iteration formula is still $x^{(k+1)} = x^{(k)} + \frac{r_i}{\|a_i\|^2} a_i$, while the probability

$$p_i = \frac{|r_i|}{\|r\|_1}. \tag{2.2.1}$$

The convergence analyses are proved as follows.

Algorithm 2.2.1 New Randomized Kaczmarz (NewRK) Algorithm

---

  procedure $(A,\ b,\ x^{(0)},\ K)$

    for $k = 0, 1, \cdots, K$ **do**

      Select $i_k \in [m]$ with probability $\Pr(\text{row}{=}i_k) = \frac{|r_{i_k}|}{\|r\|_1}$

      Set $x^{(k+1)} = x^{(k)} + \frac{b_{i_k} - \langle a_{i_k}, x^{(k)} \rangle}{\|a_{i_k}\|_2^2} a_{i_k}$

    end for

    Output: $x^{(K+1)}$

  end procedure

Lemma 2.2.1. *(Chebyshev's sum inequality) If* $a = \{a_1, a_2, \cdots, a_m\}$ *and* $b = \{b_1, b_2, \cdots, b_m\}$ *are two real numbers sequences such that* $p_1 \geqslant p_2 \geqslant$

$\cdots \geqslant p_m$ and $q_1 \geqslant q_2 \geqslant \cdots \geqslant q_m$, then the following inequality is true

$$\frac{1}{m} \sum_{k=1}^{m} p_k \sum_{k=1}^{m} q_k \leqslant \sum_{k=1}^{m} p_k q_k. \tag{2.2.2}$$

Similarly, if $p_1 \leqslant p_2 \leqslant \cdots \leqslant p_m$ and $q_1 \geqslant q_2 \geqslant \cdots \geqslant q_m$, then

$$\sum_{k=1}^{m} p_k q_k \leqslant \frac{1}{m} \sum_{k=1}^{m} p_k \sum_{k=1}^{m} q_k. \tag{2.2.3}$$

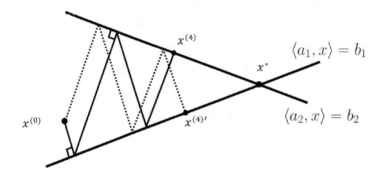

Figure 2.2.1: NewRK, $m = 2$

Theorem 2.2.1. ([43]) Let x^* be the solution of the full rank consistent system (1.1.1), if $\|a_i\| = \beta$, then NewRK linearly converges to x^* in expectation, with the average error

$$E\|x^{(k)} - x^*\|^2 \leqslant \left(1 - \kappa_F^{-2}(A)\right)^k \|x^{(0)} - x^*\|^2, \tag{2.2.4}$$

where $\kappa_F(A) = \|A^+\|_2 \|A\|_F$.

Proof. Similar to the proof of convergence of RK. Let the orthogonal projection of the kth iteration P_k satisfy

$$P_k x^{(k-1)} = x^{(k-1)} - \left\langle x^{(k-1)} - x^*, Z_{k-1} \right\rangle Z_{k-1},$$

where $Z_i = \frac{a_i}{\|a_i\|_2}$. By the orthogonality of $x^{(k-1)} - x^{(k)}$ and $x^{(k)} - x^*$,

$$
\begin{aligned}
\|x^{(k)} - x^*\|^2 &= \|x^{(k-1)} - x^*\|^2 - \|x^{(k)} - x^{(k-1)}\|^2 \\
&= \|x^{(k-1)} - x^*\|^2 - |\langle x^{(k-1)} - x^*, Z_k\rangle|^2 \\
&= \|x^{(k-1)} - x^*\|^2 - \left|\left\langle \frac{x^{(k-1)} - x^*}{\|x^{(k-1)} - x^*\|}, Z_k\right\rangle\right|^2 \|x^{(k-1)} - x^*\|^2 \\
&= \left(1 - \left|\left\langle \frac{x^{(k-1)} - x^*}{\|x^{(k-1)} - x^*\|}, Z_k\right\rangle\right|^2\right) \|x^{(k-1)} - x^*\|^2.
\end{aligned}
$$

$$(2.2.5)$$

Now we take the expectation of both sides. Then,

$$
\begin{aligned}
& E_{\{Z_1,\cdots,Z_{k-1}\}} \|x^{(k)} - x^*\|^2 \\
& \leqslant \left(1 - E_{\{Z_1,\cdots,Z_{k-1}\}} \left|\left\langle \frac{x^{(k-1)} - x^*}{\|x^{(k-1)} - x^*\|}, Z_k\right\rangle\right|^2\right) \|x^{(k-1)} - x^*\|^2.
\end{aligned}
$$

$$(2.2.6)$$

By the independence of the projection and Lemma 2.2.1,

$$
\begin{aligned}
& E_{\{Z_1,\cdots,Z_{k-1}\}} \left|\left\langle \frac{x^{(k-1)} - x^*}{\|x^{(k-1)} - x^*\|}, Z_k\right\rangle\right|^2 \\
&= \sum_{i=1}^{m} \frac{|r_i|}{\|r\|_1} \left|\left\langle \frac{x^{(k-1)} - x^*}{\|x^{(k-1)} - x^*\|}, \frac{a_i}{\|a_i\|}\right\rangle\right|^2 \\
&= \frac{1}{\beta^2 \|x^{(k-1)} - x^*\|^2} \sum_{i=1}^{m} \frac{|r_i|}{\|r\|_1} |r_i|^2 \\
&\geqslant \frac{\sum_{i=1}^{m} |r_i|^2}{m\beta^2 \|x^{(k-1)} - x^*\|^2} \\
&= \sum_{i=1}^{m} \frac{\|a_i\|^2}{\|A\|_F^2} \left|\left\langle \frac{x^{(k-1)} - x^*}{\|x^{(k-1)} - x^*\|}, \frac{a_i}{\|a_i\|}\right\rangle\right|^2 \\
&\geqslant \kappa_F^{-2}(A).
\end{aligned}
$$

$$(2.2.7)$$

Combine (2.2.6) and (2.2.7),

$$
E_{\{Z_1,\cdots,Z_{k-1}\}} \|x^{(k)} - x^*\|^2 \leqslant \left(1 - \kappa_F^{-2}(A)\right) \|x^{(k-1)} - x^*\|^2. \qquad (2.2.8)
$$

Taking the full expectation of both sides, we conclude that

$$E\|x^{(k)} - x^*\|^2 \leqslant \left(1 - \kappa_F^{-2}(A)\right) E\|x^{(k-1)} - x^*\|^2.$$

Thus we have proved that the NewRK method is as good as the RK method at least. □

Remark 2.2.1. *In fact, from (2.2.7) we can prove that*

$$E|\langle z, Z_k \rangle|^2 = \sum_{i=1}^{m} \frac{|r_i|}{\|r\|_1} \left|\left\langle z, \frac{a_i}{\|a_i\|} \right\rangle\right|^2 \geqslant \frac{\sigma_{min}(\Sigma A)}{m} \geqslant \frac{\sigma_{min}(A)}{m \max_i \|a_i\|},$$

where $\Sigma = diag(1/\|a_1\|, \cdots, 1/\|a_m\|)$.

Probability p_i can also be chosen as follows (see next section)

$$p_i = \frac{r_i^2}{\|r\|_2^2}.$$

Remark 2.2.2. *Consider an inequality problem: for $0 \leqslant a_1, a_2, \cdots, a_m$, let*

$$M_\alpha = \sum_{i=1}^{m} \alpha_i a_i, \qquad M_\beta = \sum_{i=1}^{m} \beta_i a_i,$$

where $0 \leqslant \alpha_i, \beta_i \leqslant 1, \sum_{i=1}^{m} \alpha_i = \sum_{i=1}^{m} \beta_i = 1$. When is $M_\alpha > M_\beta$ established? At least, the NewRK method's probability $\frac{|r_i|}{\|r\|_1}$ is in-order with $\left|\left\langle \frac{x^{(k-1)} - x^*}{\|x^{(k-1)} - x^*\|}, \frac{a_i}{\|a_i\|} \right\rangle\right|^2$, but the RK method's probability $\frac{\|a_i\|^2}{\|A\|_F^2}$ is not sure in-order with it.

Remark 2.2.3. *In mathematics, the **rearrangement inequality** states that*

$$x_n y_1 + \cdots + x_1 y_n \leqslant x_{\sigma(1)} y_1 + \cdots + x_{\sigma(n)} y_n \leqslant x_1 y_1 + \cdots + x_n y_n$$

for every choice of real numbers $x_1 \leqslant \cdots \leqslant x_n$ and $y_1 \leqslant \cdots \leqslant y_n$ and every permutation $x_{\sigma(1)}, \ldots, x_{\sigma(n)}$ of $x_1, \cdots, x_n$.

If the numbers are different, meaning that

$$x_1 < \cdots < x_n \quad \text{and} \quad y_1 < \cdots < y_n,$$

then the lower bound of $\sum\limits_{i=1}^{n} x_{\sigma(i)} y_i$ is attained only for the permutation which reverses the order, i.e. $\sigma(i) = n - i + 1$ for all $i = 1, \cdots, n$, and the upper bound of it is attained only for the identity, i.e. $\sigma(i) = i$ for all $i = 1, \cdots, n$.

Note that the rearrangement inequality makes no assumptions on the signs of the real numbers. Many important inequalities can be proved by the rearrangement inequality, such as **the arithmetic mean-geometric mean inequality, the Cauchy-Schwarz inequality**, and **Chebyshev's sum inequality**.

Remark 2.2.4. *As can be seen from* **Remark 2.2.2**, *let* $a_q = \max\limits_{1 \leqslant i \leqslant m} a_i$, $\alpha_q = 1$, $\alpha_i = 0$ $(i = 1, 2, \cdots, q-1, q+1, \cdots, m)$, *then* M_α *gets its maximum value. This method becomes the MCK method (see section 1.3).*

2.3 New Random Kaczmarz Method for $p_i = \frac{r_i^2}{\|r\|^2}$

In last section, the new randomized Kaczmarz method (NewRK) proposed by Guo Junhan and Li Weiguo[43] in which the row of A is chosen with probability proportional to absolute of the residual. In this section, we will prove that a new method[44], which with probability proportional to the square of the residual, converges to the least-norm solution of the linear system when it is consistent.

Observe the iterative process of the RK method, and it holds that

$$\|x^{(k+1)} - x^*\|^2 = \|x^{(k)} - x^*\|^2 - \frac{(\langle a_{i_k}, x^{(k)}\rangle - b_{i_k})^2}{\|a_{i_k}\|^2}.$$

This tells us if we choose the row of A such that $\frac{(\langle a_{i_k}, x^{(k)}\rangle - b_{i_k})^2}{\|a_{i_k}\|_2^2}$ is relatively larger than that of the other rows, better progress in each iteration can be obtained. Intuitively, we want to make the current iteration point preferentially project to a further hyperplane. With this observation, we construct an NewRK2 method which selects the row of coefficient matrix A with probability

$$p_{i_k} = \frac{|r_{i_k}|^2}{\|r\|^2},$$

and the algorithm takes the form

$$x^{(k+1)} = x^{(k)} + \frac{b_{i_k} - \langle a_{i_k}, x^{(k)} \rangle}{\|a_{i_k}\|^2} a_{i_k}.$$

Note that the most expensive component at the iteration k of the new randomized Kaczmarz method is computing the residual vector $r^{(k)}$. From the definition of the NewRK2 method, it holds that

$$
\begin{aligned}
r^{(k+1)} &= b - Ax^{(k+1)} \\
&= b - A\left(x^{(k)} + \frac{b_{i_k} - \langle a_{i_k}, x^{(k)} \rangle}{\|a_{i_k}\|^2} a_{i_k}\right) \\
&= b - A\left(x^{(k)} + \frac{r_{i_k}^{(k)}}{\|a_{i_k}\|^2} a_{i_k}\right) \\
&= r^{(k)} - \frac{r_{i_k}^{(k)}}{\|a_{i_k}\|^2} A a_{i_k} \\
&= r^{(k)} - \frac{r_{i_k}^{(k)}}{B_{i_k i_k}} B_{i_k}
\end{aligned}
$$

where $B = AA^T$, B_{i_k} is the i_kth column of B and $B_{i_k i_k}$ is the (i_k, i_k) entry of B. If we compute the matrix B at the beginning, the NewRK2 method could be faster. Based on the above considerations, we now give the NewRK2 method as follows.

In fact, if the new randomized Kaczmarz method is convergent, then it must converge to the least-norm solution $x^* = A^+ b$ of the linear system (1.1.1), provided that the initial point $x^{(0)} \in \mathcal{R}(A^T)$. From the consistency of the iteration scheme, any convergent point $\tilde{x}$ of the new randomized Kaczmarz iteration sequence should satisfy the equations

$$(b_i - \langle a_i, \tilde{x} \rangle) a_i^T = 0, i = 1, 2, \ldots, m.$$

Thus, it holds that

$$A^T b - A^T A\tilde{x} = \sum_{i=1}^{m} (b_i - \langle a_i, \tilde{x} \rangle) a_i^T = 0,$$

Algorithm 2.3.1 New Randomized Kaczmarz (NewRK2) Method

---

procedure $(A, \ b, \ x^{(0)} \in \mathcal{R}(A^T), \ K, \ B = AA^T)$

    for $k = 0, 1, \cdots, K$ **do**

        Select an index $i_k \in \{1, 2, \ldots, m\}$ with probability $p_{i_k} = \frac{|r_{i_k}|^2}{\|r\|^2}$

        Compute $x^{(k+1)} = x^{(k)} + \frac{r_{i_k}}{B_{i_k i_k}} a_{i_k}$

        Compute $r^{(k+1)} = r^{(k)} - \frac{r_{i_k}}{B_{i_k i_k}} B_{i_k}$

    end for

    Output: $x^{(K+1)}$

end procedure

---

that is,

$$A^T b = A^T A \tilde{x}.$$

Note that the $\tilde{x}$ can be written as

$$\tilde{x} = A^+ b + (I - P_{A^T}) z, \quad z \in \mathcal{R}^n$$

where P_{A^T} is the orthogonal projection onto the range space of A^T. Since $x^{(0)} \in \mathcal{R}(A^T)$, the new randomized Kaczmarz iteration only adds a linear combination of a column of A^T to the current iterate at each step, $\tilde{x} \in \mathcal{R}(A^T)$, so that $\tilde{x} \equiv x^* = A^+ b$.

To prove the convergence of the NewRK2, we give two lemmas as follows.

Lemma 2.3.1. (Jensen inequality) *Let $I \in \mathcal{R}$ be an interval. If function $f : I \to \mathcal{R}$ is strictly convex, then*

$$f\left(\sum_{i=1}^{n} t_i x_i\right) \leqslant \sum_{i=1}^{n} t_i f(x_i), \tag{2.3.1}$$

for all $n \in N^+, x_1, \ldots, x_n \in I$ and $t_1, \ldots, t_n > 0$ with $t_1 + \ldots + t_n = 1$. If and only if $x_1 = \ldots = x_n$, the equation holds.

Lemma 2.3.2. *The following inequality holds*

$$\frac{a^{\eta+1}}{x^\eta} + \frac{b^{\eta+1}}{y^\eta} \geqslant \frac{(a+b)^{\eta+1}}{(x+y)^\eta} \tag{2.3.2}$$

for $a \geqslant 0, b \geqslant 0, x > 0, y > 0, \eta > 0$.

Proof. We first consider the case $a \neq 0$, $b \neq 0$. Let $\alpha = \frac{a}{a+b} \neq 0, \beta = \frac{b}{a+b} \neq 0$, from (2.3.2) we derive

$$\alpha \frac{1}{\left(\frac{x}{\alpha}\right)^\eta} + \beta \frac{1}{\left(\frac{y}{\beta}\right)^\eta} \geqslant \frac{1}{(x+y)^\eta}. \tag{2.3.3}$$

Take function $f : I \subseteq (0, \infty) \to R$ of the form $f(x) = \frac{1}{x^\eta}$. Since $f''(x) = \frac{\eta(\eta+1)}{x^{\eta+2}} > 0$, the $f(x)$ is strictly convex. Let $x_1 = \frac{x}{\alpha} > 0, y_1 = \frac{y}{\beta} > 0$, by Lemma 2.3.1 we have the following inequality

$$\alpha \frac{1}{x_1^\eta} + \beta \frac{1}{y_1^\eta} \geqslant \frac{1}{(\alpha x_1 + \beta y_1)^\eta},$$

Then (2.3) follows.

Obviously, it also holds when $a = 0$ or $b = 0$. The proof is complete. $\quad\square$

Remark 2.3.1. *Note that $f(x) = \frac{1}{x^\eta} (x > 0, \eta > 0)$ is strictly convex, the equation (2.3.2) holds if and only if $\frac{a}{x} = \frac{b}{y}$.*

With the use of Lemma 2.3.2 and induction, it is easy to derive the following corollary.

Corollary 2.3.1. *The following inequality holds*

$$\sum_{i=1}^{m} \frac{x_i^{\eta+1}}{y_i^\eta} \geqslant \frac{\left(\sum\limits_{i=1}^{m} x_i\right)^{\eta+1}}{\left(\sum\limits_{i=1}^{m} y_i\right)^\eta}. \tag{2.3.4}$$

for $m \in N^+, \eta > 0, x_i \geqslant 0, y_i > 0, i = 1, 2, \ldots, m$.

Remark 2.3.2. *Note that $f(x) = \frac{1}{x^\eta} (x > 0, \eta > 0)$ is strictly convex, so the equation (2.3.4) holds if and only if $\frac{x_1}{y_1} = \ldots = \frac{x_m}{y_m}$.*

For the convergence property of the NewRK2, we can establish the following theorem.

Theorem 2.3.1. *Let the linear system (1.1.1) be consistent. Then the iteration sequence* $\left\{x^{(k)}\right\}_{k=0}^{\infty}$ *generated by the new randomized Kaczmarz method (NewRK2) staring from any initial approximation* $x^{(0)} \in \mathcal{R}(A^T)$, *converges to the least-norm solution* $x^* = A^+ b$ *in expectation. The solution error in expectation for the iteration sequence* $\left\{x^{(k)}\right\}_{k=0}^{\infty}$ *obeys*

$$E \left\|x^{(1)} - x^*\right\|^2 \leqslant \left(1 - \frac{\lambda_{\min}(A^T A)}{\|A\|_F^2}\right) \left\|x^{(0)} - x^*\right\|^2 \qquad (2.3.5)$$

and

$$E_k \left\|x^{(k+1)} - x^*\right\|^2 \leqslant \left(1 - \frac{\lambda_{\min}(A^T A)}{\gamma}\right) \left\|x^{(k)} - x^*\right\|^2, k = 1, 2, \cdots.$$
$$(2.3.6)$$

As a result, it holds that

$$E \left\|x^{(k)} - x^*\right\|^2 \leqslant \left(1 - \frac{\lambda_{\min}(A^T A)}{\gamma}\right)^{k-1} \left(1 - \frac{\lambda_{\min}(A^T A)}{\|A\|_F^2}\right) \left\|x^{(0)} - x^*\right\|^2,$$
$$(2.3.7)$$

where $\gamma = \|A\|_F^2 - \min\limits_{1 \leqslant i \leqslant m} \|a_i\|^2$, $\lambda_{\min}(A^T A)$ *is the smallest nonzero eigenvalues of* $A^T A$.

Proof. From the definition of the new randomized Kaczmarz method, for $k = 1, 2, \ldots$, we have

$$
\begin{aligned}
r_{i_{k-1}}^{(k)} &= b_{i_{k-1}} - \langle a_{i_{k-1}}, x^{(k)} \rangle \\
&= b_{i_{k-1}} - a_{i_{k-1}} \left(x^{(k-1)} + \frac{b_{i_{k-1}} - \langle a_{i_{k-1}}, x^{(k-1)} \rangle}{\|a_{i_{k-1}}\|^2} a_{i_{k-1}}\right) \\
&= b_{i_{k-1}} - \langle a_{i_{k-1}}, x^{(k-1)} \rangle - \left(b_{i_{k-1}} - \langle a_{i_{k-1}}, x^{(k-1)} \rangle\right) \\
&= 0.
\end{aligned}
\qquad (2.3.8)
$$

In addition, we have

$$x^{(k+1)} - x^{(k)} = \frac{\left(b_{i_k} - \langle a_{i_k}, x^{(k)} \rangle\right)}{\|a_{i_k}\|^2} a_{i_k},$$

which implies that $x^{(k+1)} - x^{(k)}$ is parallel to a_{i_k}. Since $\langle a_{i_k}, x^* \rangle = b_{i_k}$, it holds that

$$
\begin{aligned}
\langle a_{i_k}, x^{(k+1)} - x^* \rangle &= \langle a_{i_k}, x^{(k)} - x^* + \tfrac{b_{i_k} - \langle a_{i_k}, x^{(k)} \rangle}{\|a\|^2} a_{i_k} \rangle \\
&= \langle a_{i_k}, x^{(k)} - x^* \rangle + \left(b_{i_k} - \langle a_{i_k}, x^{(k)} \rangle \right) \\
&= b_{i_k} - \langle a_{i_k}, x^* \rangle \\
&= 0,
\end{aligned}
$$

which shows that $x^{(k+1)} - x^*$ is orthogonal to a_{i_k}. Therefore, $x^{(k+1)} - x^*$ is perpendicular to $x^{(k+1)} - x^{(k)}$, i.e. $\langle x^{(k+1)} - x^{(k)}, x^{(k+1)} - x^* \rangle = 0$. It holds that

$$
\begin{aligned}
\left\| x^{(k+1)} - x^* \right\|^2 &= \left\| x^{(k)} - x^* \right\|^2 - \left\| x^{(k+1)} - x^{(k)} \right\|^2 \\
&= \left\| x^{(k)} - x^* \right\|^2 - \left| \langle x^{(k)} - x^*, Z_{i_k} \rangle \right|^2 \\
&= \left\| x^{(k)} - x^* \right\|^2 - \left| \langle \tfrac{x^{(k)} - x^*}{\|x^{(k)} - x^*\|}, Z_{i_k} \rangle \right|^2 \left\| x^{(k)} - x^* \right\|^2 \\
&= \left(1 - \left| \langle \tfrac{x^{(k)} - x^*}{\|x^{(k)} - x^*\|}, Z_{i_k} \rangle \right|^2 \right) \left\| x^{(k)} - x^* \right\|^2,
\end{aligned}
$$

$$(2.3.9)$$

where $Z_{i_k} = \frac{a_{i_k}}{\|a_{i_k}\|}$ is a random vector chosen with probability $\frac{\left| r_{i_k}^{(k)} \right|^2}{\left\| r^{(k)} \right\|^2}$.

Based on (2.3.9), by taking the conditional expectation on both sides, it holds that

$$
\begin{aligned}
&E_k \left\| x^{(k+1)} - x^* \right\|^2 \\
&= \left(1 - E_k \left| \langle \tfrac{x^{(k)} - x^*}{\|x^{(k)} - x^*\|}, Z_{i_k} \rangle \right|^2 \right) \left\| x^{(k)} - x^* \right\|^2.
\end{aligned} \qquad (2.3.10)
$$

By the definition of the expectation and the independent realization of the random projection, for $k = 1, 2, \ldots$, we have

$$
\begin{aligned}
E_k \left| \left\langle \frac{x^{(k)} - x^*}{\|x^{(k)} - x^*\|}, Z_{i_k} \right\rangle \right|^2
&= \sum_{i_k=1}^{m} \frac{\left| r_{i_k}^{(k)} \right|^2}{\|r^{(k)}\|^2} \left| \left\langle \frac{x^{(k)} - x^*}{\|x^{(k)} - x^*\|}, \frac{a_{i_k}}{\|a_{i_k}\|} \right\rangle \right|^2 \\
&= \sum_{\substack{i_k=1 \\ i_k \neq i_{k-1}}}^{m} \frac{\left| r_{i_k}^{(k)} \right|^2}{\|r^{(k)}\|^2} \left| \left\langle \frac{x^{(k)} - x^*}{\|x^{(k)} - x^*\|}, \frac{a_{i_k}}{\|a_{i_k}\|} \right\rangle \right|^2 \\
&= \frac{1}{\|r^{(k)}\|^2} \frac{1}{\|x^{(k)} - x^*\|^2} \sum_{\substack{i_k=1 \\ i_k \neq i_{k-1}}}^{m} \frac{\left(\left| r_{i_k}^{(k)} \right|^2 \right)^2}{\|a_{i_k}\|^2} \\
&\geq \frac{1}{\|r^{(k)}\|^2} \frac{1}{\|x^{(k)} - x^*\|^2} \frac{\left(\displaystyle\sum_{\substack{i_k=1 \\ i_k \neq i_{k-1}}}^{m} \left| r_{i_k}^{(k)} \right|^2 \right)^2}{\displaystyle\sum_{\substack{i_k=1 \\ i_k \neq i_{k-1}}}^{m} \|a_{i_k}\|^2} \\
&\geq \frac{1}{\|r^{(k)}\|^2} \frac{1}{\|x^{(k)} - x^*\|^2} \frac{\|r^{(k)}\|^2 \displaystyle\sum_{i_k=1}^{m} \left| r_{i_k}^{(k)} \right|^2}{\gamma} \\
&= \frac{1}{\gamma} \sum_{i_k=1}^{m} \left| \left\langle \frac{x^{(k)} - x^*}{\|x^{(k)} - x^*\|}, a_{i_k} \right\rangle \right|^2 \\
&\geq \frac{\lambda_{\min}(A^T A)}{\gamma}.
\end{aligned}
\tag{2.3.11}
$$

Here, the second equality is valid due to $r_{i_{k-1}}^{(k)} = 0$ (see (2.3.8)). The first inequality is achieved with the use of Corollary 2.3.1, where $\eta = 1$. For the last inequality, the following estimate

$$
\|Az\|^2 \geq \lambda_{\min}(A^T A)\|z\|^2,
$$

is used, where $z \in \mathcal{R}(A^T)$. For $k = 0$, similar to (2.3.11), we can get

$$
E \left| \left\langle \frac{x^{(0)} - x^*}{\|x^{(0)} - x^*\|}, Z_{i_0} \right\rangle \right|^2 \geq \frac{\lambda_{\min}(A^T A)}{\|A\|_F^2}.
\tag{2.3.12}
$$

Then we can obtain an estimate as follows:

$$
E_k \left\| x^{(k+1)} - x^* \right\|^2 \leq \left(1 - \frac{\lambda_{\min}(A^T A)}{\gamma} \right) \left\| x^{(k)} - x^* \right\|^2, \quad k = 1, 2, \ldots,
\tag{2.3.13}
$$

and

$$E\left\|x^{(1)} - x^*\right\|^2 \leqslant \left(1 - \frac{\lambda_{\min}(A^T A)}{\|A\|_F^2}\right)\left\|x^{(0)} - x^*\right\|^2. \tag{2.3.14}$$

Finally, by taking the full expectation on both sides of (2.3.13), we have

$$E\left\|x^{(k+1)} - x^*\right\|^2 \leqslant \left(1 - \frac{\lambda_{\min}(A^T A)}{\gamma}\right) E\left\|x^{(k)} - x^*\right\|^2, k = 1, 2, \ldots, \tag{2.3.15}$$

By induction on the iteration index k, we straightforwardly obtain the estimate (2.3.7). □

Remark 2.3.3. *Theorem 2.3.1 shows that the NewRK2 method should converge significantly faster than the RK method. In addition, set $d_i = \frac{|r_i|}{\|a_i\|}$, $i = 1, 2, \ldots, m$, the equality (2.3.11) holds if and only if $d_1 = d_2 = \ldots = d_m$. Thus the inequality holds strictly (2.3.11) unless $d_1 = d_2 = \ldots d_m$.*

Let $\rho_{RK} = 1 - \frac{\lambda_{\min}(A^T A)}{\|A\|_F^2}$, $\rho_{NewRK2} = 1 - \frac{\lambda_{\min}(A^T A)}{\gamma}$, it is easy to get the following conclusion.

Corollary 2.3.2. *If A has no zero row, then the following inequality holds.*

$$\rho_{RK} - \rho_{NewRK2} \geqslant \frac{\min\limits_{1\leqslant i\leqslant m}\|a_i\|^2}{\|A\|_F^2}\frac{1}{\kappa^2(A)} > 0. \tag{2.3.16}$$

Especially, $\rho_{RK} - \rho_{NewRK2} \geqslant \frac{1}{m}\frac{1}{\kappa^2(A)} > 0$ when $\|a_1\| = \|a_2\| = \ldots = \|a_m\|$.

Proof. Obviously,

$$\begin{aligned}
\rho_{RK} - \rho_{NewRK2} &= \frac{\min\limits_{1\leqslant i\leqslant m}\|a_i\|^2 \cdot \lambda_{\min}(A^T A)}{\gamma \cdot \|A\|_F^2} \\
&\geqslant \frac{\min\limits_{1\leqslant i\leqslant m}\|a_i\|^2}{\|A\|_F^2} \cdot \frac{\lambda_{\min}(A^T A)}{\|A\|_F^2} \\
&> 0.
\end{aligned}$$

From $\|A\|_F^2 = \sum\limits_{i=1}^{m}\|a_i\|^2$, and $\kappa_F^2(A) = \frac{\|A\|_F^2}{\lambda_{\min}(A^T A)}$, we get the corollary. □

Remark 2.3.4. *From the discussion in these two sections, we can see that the convergence rate of the random Kaczmarz method, in which the row of A is chosen with probability proportional to $|r_i|^2$, is significantly better than that of the random Kaczmarz method, in which the row of A is chosen with probability proportional to $|r_i|$. In fact, if the row of A is chosen with probability proportional to $|r_i|^p$, it can be observed numerically that larger values of p result in approximations for which $\|Ax^{(k)} - b\|_\infty$ decays more rapidly. For the theoretical analysis of this fact(see [45]).*

2.4   Greedy Randomized Kaczmarz Method

In 2018, an effective probability criterion for selecting the working rows from the coefficient matrix to construct a greedy randomized Kaczmarz method was introduced[36]. This method converges to the unique least-norm solution of the linear system when it is consistent. Theoretical analysis demonstrates that the convergence rate of the greedy randomized Kaczmarz method is much faster than the randomized Kaczmarz method, and numerical results also show that the greedy randomized Kaczmarz method is more efficient than the randomized Kaczmarz method.

In fact, at the kth iterate, for the corresponding residual vector $r^{(k)} = b - Ax^{(k)}$, if $|r_i^{(k)}| > |r_j^{(k)}|$, $i, j \in [m]$, we may want the ith row to be selected with a larger probability than the jth row, so that larger entries of the residual vector $r^{(k)}$ can be preferentially annihilated as far as possible. In this manner, the correspondingly induced randomized Kaczmarz method should converge to the solution of the linear system (1.1.1) with a fast convergence rate. Motivated by this observation, the greedy randomized Kaczmarz method is constructed as Algorithm 2.4.1.

The greedy randomized Kaczmarz method is well defined as the index set $\mathcal{U}_k$ defined in the following algorithm is nonempty for all iteration index

---

Algorithm 2.4.1 Greedy Randomized Kaczmarz (GRK) Algorithm

---

procedure $(A,\ b,\ x^{(0)},\ K)$

    for $k = 0, 1, \cdots, K$ **do**

        Compute

$$\epsilon_k = \frac{1}{2}\left(\frac{1}{\|b - Ax^{(k)}\|^2}\max_{1 \leqslant i_k \leqslant m}\left\{\frac{|b_{i_k} - \langle a_i, x^{(k)}\rangle|^2}{\|a_{i_k}\|^2}\right\} + \frac{1}{\|A\|_F^2}\right) \qquad (2.4.1)$$

        Determine the index set of positive integers

$$\mathcal{U}_k = \left\{i_k \big| b_{i_k} - \langle a_i, x^{(k)}\rangle|^2 \geqslant \epsilon_k \|b - Ax^{(k)}\|^2 \|a_{i_k}\|^2\right\} \qquad (2.4.2)$$

        Compute the *ith* entry $\tilde{r}_i^{(k)}$ of the vector $\tilde{r}^{(k)}$ according to

$$\tilde{r}_i^{(k)} = \begin{cases} b_i - \langle a_i, x^{(k)}\rangle, & \text{if } i \in \mathcal{U}_k, \\ 0, & \text{otherwise.} \end{cases}$$

        Select $i_k \in \mathcal{U}_k$ with probability $Pr(row = i_k) = \dfrac{|\tilde{r}_{i_k}^{(k)}|^2}{\|\tilde{r}^{(k)}\|^2}$

        Set $x^{(k+1)} = x^{(k)} + \dfrac{b_{i_k} - \langle a_i, x^{(k)}\rangle}{\|a_{i_k}\|^2}a_{i_k}$

    end for

    Output: $x^{(K+1)}$

end procedure

---

k. Indeed, we can show that $i \in \mathcal{U}_k$ for $k \in \{0, 1, 2, \cdots\}$ if

$$\frac{|b_i - \langle a_i, x^{(k)} \rangle|^2}{\|a_i\|^2} = \max_{1 \leqslant i_k \leqslant m} \left(\frac{|b_{i_k} - \langle a_i, x^{(k)} \rangle|^2}{\|a_{i_k}\|^2} \right).$$

This is because of

$$\max_{1 \leqslant i_k \leqslant m} \left(\frac{|b_{i_k} - \langle a_i, x^{(k)} \rangle|^2}{\|a_{i_k}\|^2} \right) \geqslant \sum_{i_k=1}^{m} \frac{\|a_{i_k}\|^2}{\|A\|_F^2} \cdot \frac{|b_{i_k} - \langle a_i, x^{(k)} \rangle|^2}{\|a_{i_k}\|^2} = \frac{\|b - Ax^{(k)}\|^2}{\|A\|_F^2}$$

For the convergence property of the greedy randomized Kaczmarz method (GRK), there exists the following theorem.

Theorem 2.4.1. *Let the linear system (1.1.1) with the coefficient matrix* $A \in \mathcal{R}^{m \times n}$ *and the right-hand side* $b \in \mathcal{R}^m$, *be consistent. Then the iteration sequence* $\{x^{(k)}\}_{k=0}^{\infty}$, *generated by the greedy randomized Kaczmarz method starting from any initial guess* $x^{(0)} \in \mathcal{R}^n$ *in the column space of* A^T, *converges to the unique least-norm solution* $x^* = A^{\dagger}b$ *in expectation. Moreover, the solution error in expectation for the iteration sequence* $\{x^{(k)}\}_{k=0}^{\infty}$ *obeys*

$$E\|x^{(1)} - x^*\|^2 \leqslant \left(1 - \frac{\lambda_{\min}(A^T A)}{\|A\|_F^2} \right) \|x^{(0)} - x^*\|^2, \qquad (2.4.3)$$

and

$$E_k\|x^{(k+1)} - x^*\|^2 \leqslant \left[1 - \frac{1}{2}(\frac{1}{\gamma}\|A\|_F^2 + 1)\frac{\lambda_{\min}(A^T A)}{\|A\|_F^2} \right] \|x^{(k)} - x^*\|^2, k = 1, 2, \cdots$$
$$(2.4.4)$$

where $\gamma = \max_{1 \leqslant i \leqslant m} \sum_{j=1, j \neq i}^{m} \|a_j\|^2$. *As a result, it holds that*

$$E\|x^{(k)} - x^*\|^2$$

$$\leqslant \left[1 - \frac{1}{2}(\frac{1}{\gamma}\|A\|_F^2 + 1)\frac{\lambda_{\min}(A^T A)}{\|A\|_F^2} \right]^{k-1} \left(1 - \frac{\lambda_{\min}(A^T A)}{\|A\|_F^2} \right) \|x^{(0)} - x^*\|^2.$$
$$(2.4.5)$$

Proof. From the definition of the greedy randomized Kaczmarz method, for

$k = 1, 2, \cdots$, we have

$$
\begin{aligned}
r_{i_{k-1}}^{(k)} &= b_{i_{k-1}} - \langle a_{i_{k-1}}, x^{(k)} \rangle \\
&= b_{i_{k-1}} - \left\langle a_{i_{k-1}}, x^{(k-1)} + \frac{b_{i_{k-1}} - \langle a_{i_{k-1}}, x^{(k-1)} \rangle}{\|a_{i_{k-1}}\|^2} a_{i_{k-1}} \right\rangle \\
&= b_{i_{k-1}} - \langle a_{i_{k-1}}, x^{(k-1)} \rangle - (b_{i_{k-1}} - \langle a_{i_{k-1}}, x^{(k-1)} \rangle) \\
&= 0.
\end{aligned}
$$

It then follows for $k = 1, 2, \cdots$ that

$$
\begin{aligned}
\epsilon_k \|A\|_F^2 &= \frac{\max\limits_{1 \leqslant i_k \leqslant m} \left(\frac{|b_{i_k} - \langle a_{i_k}, x^{(k)} \rangle|^2}{\|a_{i_k}\|^2} \right)}{2 \sum\limits_{i_k=1}^{m} \frac{\|a_{i_k}\|^2}{\|A\|_F^2} \cdot \frac{|b_{i_k} - \langle a_{i_k}, x^{(k)} \rangle|^2}{\|a_{i_k}\|^2}} + \frac{1}{2} \\
&= \frac{\max\limits_{1 \leqslant i_k \leqslant m} \left(\frac{|b_{i_k} - \langle a_{i_k}, x^{(k)} \rangle|^2}{\|a_{i_k}\|^2} \right)}{2 \sum\limits_{\substack{i_k=1 \\ i_k \neq i_{k-1}}}^{m} \frac{\|a_{i_k}\|^2}{\|A\|_F^2} \cdot \frac{|b_{i_k} - \langle a_{i_k}, x^{(k)} \rangle|^2}{\|a_{i_k}\|^2}} + \frac{1}{2} \\
&\geqslant \frac{1}{2} \left(\frac{\|A\|_F^2}{\sum\limits_{\substack{i_k=1 \\ i_k \neq i_{k-1}}}^{m} \|a_{i_k}\|^2} + 1 \right) \\
&\geqslant \frac{1}{2} \left(\frac{1}{\gamma} \|A\|_F^2 + 1 \right).
\end{aligned}
$$

For $k = 0$, we can get

$$
\begin{aligned}
\epsilon_0 \|A\|_F^2 &= \frac{\max\limits_{1 \leqslant i_k \leqslant m} \left(\frac{|b_{i_0} - \langle a_{i_0}, x^{(0)} \rangle|^2}{\|a_{i_0}\|^2} \right)}{2 \sum\limits_{i_0=1}^{m} \frac{\|a_{i_0}\|^2}{\|A\|_F^2} \cdot \frac{|b_{i_0} - \langle a_{i_0}, x^{(0)} \rangle|^2}{\|a_{i_0}\|^2}} + \frac{1}{2} \\
&\geqslant \frac{1}{2} + \frac{1}{2} \\
&= 1.
\end{aligned}
$$

Again, from the definition of the greedy randomized Kaczmarz method we obtain

$$x^{(k+1)} - x^{(k)} = \frac{b_{i_k} - \langle a_{i_k}, x^{(k)} \rangle}{\|a_{i_k}\|^2} a_{i_k},$$

which implies that $x^{(k+1)} - x^{(k)}$ is parallel to a_{i_k}. Since $\langle a_{i_k}, x^* \rangle = b_{i_k}$, it holds that

$$
\begin{aligned}
\langle a_{i_k}, (x^{(k+1)} - x^*) \rangle &= \left\langle a_{i_k}, x^{(k)} - x^* + \frac{b_{i_k} - a_{i_k} x^{(k)}}{\|a_{i_k}\|^2} a_{i_k} \right\rangle \\
&= \langle a_{i_k}, (x^{(k)} - x^*) \rangle + (b_{i_k} - \langle a_{i_k}, x^{(k)} \rangle) \\
&= b_{i_k} - a_{i_k} x^* \\
&= 0,
\end{aligned}
$$

which shows that $x^{(k+1)} - x^*$ is orthogonal to a_{i_k}. Therefore, we know that $x^{(k+1)} - x^*$ is perpendicular to $x^{(k+1)} - x^{(k)}$ for any $k \geqslant 0$ or, in other words, $\langle x^{(k+1)} - x^*, x^{(k+1)} - x^{(k)} \rangle = 0$. It follows that

$$\|x^{(k+1)} - x^*\|^2 = \|x^{(k)} - x^*\|^2 - \|x^{(k+1)} - x^{(k)}\|^2.$$

Based on this equality, we have

$$
\begin{aligned}
E_k \|x^{(k+1)} - x^*\|^2 &= \|x^{(k)} - x^*\|^2 - E_k \|x^{(k+1)} - x^k\|^2 \\
&= \|x^{(k)} - x^*\|^2 - \\
&\quad \sum_{i_k \in \mathcal{U}_k} \frac{|b_{i_k} - \langle a_{i_k}, x^{(k)} \rangle|^2}{\sum_{i_k \in \mathcal{U}_k} |b_{i_k} - \langle a_{i_k}, x^{(k)} \rangle|^2} \cdot \frac{|\langle a_{i_k}, x^{(k)} - x^* \rangle|^2}{\|a_{i_k}\|^2} \\
&\leqslant \|x^{(k)} - x^*\|^2 - \epsilon_k \|b - Ax^{(k)}\|^2.
\end{aligned}
$$

The last inequality is achieved with the use of the definition of ϵ which lead to

$$|b_{i_k} - \langle a_{i_k}, x^{(k)} \rangle|^2 \geqslant \epsilon_k \|b - Ax^{(k)}\|^2 \|a_{i_k}\|^2, \quad \forall i_k \in \mathcal{U}_k.$$

As $x^* = A^\dagger b$ belongs to the column space of A^T, by starting from an arbitrary initial guess $x^{(0)}$ in the column space of A^T, each step in the greedy randomized Kaczmarz iteration only adds a linear combination of a column of A^T. Hence, $x^{(k)} - x^*$ is in the column space of A^T, too. We can then obtain an estimate as follows:

$$
\begin{aligned}
E_k \| x^{(k+1)} - x^* \|^2 &\leqslant \| x^{(k)} - x^* \|^2 - \epsilon_k \| b - A x^{(k)} \|^2 \\
&= \| x^{(k)} - x^* \|^2 - \epsilon_k \| A(x^* - x^{(k)}) \|^2 \\
&\leqslant (1 - \epsilon_k \lambda_{\min}(A^T A)) \| x^{(k)} - x^* \|^2 .
\end{aligned}
$$

Here in the last inequality we have used the estimate

$$
\| A u \|_2^2 \geqslant \lambda_{\min}(A^T A)) \| u \|^2 , \tag{2.4.6}
$$

which holds true for any $u \in R^n$ belonging to the column space of A^T. By making use of

$$
\epsilon_0 \geqslant \frac{1}{\| A \|_F^2}
$$

and

$$
\epsilon_k \geqslant \frac{1}{2} \left(\frac{1}{\gamma} \| A \|_F^2 + 1 \right) \frac{1}{\| A \|_F^2} . \ k = 1, 2, \cdots ,
$$

we can obtain the estimate (2.4.3) and (2.4.4) about the convergence rate of the greedy randomized Kaczmarz iteration sequence.

Finally, by taking the full expectation on both sides of (2.4.4), we have

$$
E_k \| x^{(k+1)} - x^* \|^2 \leqslant \left[1 - \frac{1}{2} \left(\frac{1}{\gamma} \| A \|_F^2 + 1 \right) \frac{1}{\| A \|_F^2} \right] E_k \| x^{(k)} - x^* \|^2 , \ k = 1, 2, \cdots .
$$

By induction on the iteration index k, we straightforwardly obtain the estimate (2.4.5). $\qquad \square$

2.5 RGS and GRGS

It was proved by Strohmer and Vershynin[39] that for overdetermined consistent systems, the randomized Kaczmarz method (**RK**) converges with expected exponential rate, independent of the number of equations in the system. RK algorithm works on the rows of A (**data points**).

Leventhal and Lewis[20] afterward proved linear convergence of randomized coordinate descent method (**RCD**) or randomized Gauss-Seidel method

(**RGS**), which instead operates on the columns of A (**features**). Later, Ma, Needell, and Ramdas[46] provided a side-by-side analysis of RK and RGS for linear systems in a variety of under- and over-constrained settings. Indeed, both RK and RGS can be viewed in two dual ways–either as variants of s-tochastic gradient descent for minimizing an appropriate objective function or as variants of randomized coordinate descent on an appropriate linear system–and to avoid confusion. By aligning with recent literature, we refer to the row-based variant as RK and the column-based variant as RGS or RCD. The advantage of such approaches is that they do not need access to the entire system but rather only individual rows (or columns) at a time. This makes them amenable in data streaming settings or when the system is so large-scale that it may not even load into memory.

RCD or RGS Algorithm

Similar to the RK algorithm, If the columns in Algorithm 1.2.1 (CD Algorithm) are selected randomly, the following nonsymmetric randomized coordinate descent method is obtained.

Algorithm 2.5.1 Randomized Coordinate Descent (RCD) Algorithm

---

   procedure $(A,\ b,\ x^{(0)},\ K,\ r^{(0)} = b - Ax^{(0)})$

      for $k = 0, 1, \cdots, K$ **do**

         Select $i \in [n]$ with $\Pr\{\text{column}=i\} = \frac{\|A_i\|^2}{\|A\|_F^2}$

         Compute $\alpha_k = \frac{\langle A_i, r^{(k)} \rangle}{\|A_i\|^2}$,

         Compute

$$x^{(k+1)} = x^{(k)} + \alpha_k e_i, \quad and \quad r^{(k+1)} = r^{(k)} - \alpha_k A_i \qquad (2.5.1)$$

      end for

      Output: $x^{(K+1)}$

   end procedure

The next result shows that the RCD or RGS algorithm would converge to a least-squares solution of (1.1.1).

Theorem 2.5.1. *Consider any linear system $Ax = b$, where the matrix A is nonzero. Define the least-squares residual and the error by*

$$f(x) = \frac{1}{2}\|Ax - b\|^2,$$

$$\delta(x) = f(x) - \min f.$$

Then the nonsymmetric randomized coordinate descent algorithm is linearly convergent in expectation to a least squares solution for the system: for each iteration $k = 0, 1, 2, \cdots$,

$$E[\delta(x^{(k+1)})|x^{(k)}] \leqslant \left(1 - \frac{\lambda_{\min}(A^T A)}{\|A\|_F^2}\right)\delta(x^{(k)}). \tag{2.5.2}$$

In particular, if A has full column rank, we have the equivalent property

$$E[\|x^{(k+1)} - x^*\|_{A^T A}^2 |x^{(k)}] \leqslant \left(1 - \frac{1}{\kappa_F^2(A)}\right)\|x^{(k)} - x^*\|_{A^T A}^2, \tag{2.5.3}$$

where $x^* = A^+ b = (A^T A)^{-1} A^T b$ is the unique least-squares solution.

Proof. Assume that $\tilde{x}$ is any least square solution of $Ax = b$. With the use of the equality $A^T A\tilde{x} = A^T b$, it is easy to prove that

$$f(x) - f(\tilde{x}) = \frac{1}{2}\|x - \tilde{x}\|_{A^T A}^2. \tag{2.5.4}$$

Note that if coordinate direction e_i is chosen during iteration k, then there exists the following equality

$$\begin{aligned} f(x^{(k+1)}) - f(\tilde{x}) &= \frac{1}{2}\|x^{(k+1)} - \tilde{x}\|_{A^T A}^2 \\ &= \frac{1}{2}\|x^{(k)} - \tilde{x}\|_{A^T A}^2 - \frac{\left((Ax^{(k)} - b)^T A_i\right)^2}{2\|A_i\|^2}. \end{aligned} \tag{2.5.5}$$

Hence,

$$\begin{aligned} E[f(x^{(k+1)})|x^{(k)}] &= f(x^{(k)}) - \sum_{i=1}^{n} \frac{\|A_i\|^2}{\|A\|_F^2} \frac{\left((Ax^{(k)} - b)^T A_i\right)^2}{2\|A_i\|^2} \\ &= f(x^{(k)}) - \frac{(A^T(b - Ax^{(k)}))^2}{2\|A\|_F^2}. \end{aligned} \tag{2.5.6}$$

Due to $A^T((b - Ax^{(k)}) = A^T A(\tilde{x} - x^{(k)})$, it is easy to see that

$$(A^T(b - Ax^{(k)}))^2 \geqslant \lambda_{\min}(A^T A)\|\tilde{x} - x^{(k)}\|_{A^T A}^2 = \lambda_{\min}(A^T A)\delta(x^{(k)}). \tag{2.5.7}$$

Combining (2.5.4), (2.5.5) and (2.5.6), we get

$$E[\delta(x^{(k+1)})|x^{(k)}] \leqslant \left(1 - \frac{\lambda_{\min}(A^T A)}{\|A\|_F^2}\right)\delta(x^{(k)}),$$

and the first result follows. If A has full column rank, the least square solution of $Ax = b$ is unique and then $\tilde{x} = x^*$. Thus the above equation provides the second result. $\qquad\square$

Remark 2.5.1. *However, RK fails to find the least square solution for an overdetermined inconsistent linear system*[47], *and RGS fails to find the least norm solution for an underdetermined consistent linear system*[46].

GRCD or GRGS Algorithm.

Similar to the **GRK** algorithm, the following GRCD algorithm[38] (or GRGS algorithm) is easy to obtain.

At the kth iterate, RGS method has the residual vector $r^{(k)} = b - Ax^{(k)}$ for (1.1.1), equivalently, the residual vector $A^T r^{(k)} = A^T b - A^T Ax^{(k)}$ for $A^T Ax = A^T b$. If $|\langle A_i, r^{(k)}\rangle| > |\langle A_j, r^{(k)}\rangle|$, $i, j \in [n]$, we may want the ith column to be selected with a lager probability than the jth column. Motivated by this observation, we set

$$\epsilon_k = \frac{1}{2}\left(\frac{1}{\|A^T r^{(k)}\|^2}\max_{1\leqslant i_k\leqslant n}\frac{|\langle A_{i_k}, r^{(k)}\rangle|^2}{\|A_{i_k}\|^2} + \frac{1}{\|A\|_F^2}\right) \qquad (2.5.8)$$

and

$$U_k = \left\{i_k \big| |\langle A_{i_k}, r^{(k)}\rangle|^2 \geqslant \epsilon_k\|A^T r^{(k)}\|^2\|A_{i_k}\|^2\right\}. \qquad (2.5.9)$$

GRGS updates $x^{(k+1)}$ in the index set U_k in each iteration.

Lemma 2.5.1. *If*

$$\frac{|\langle A_i, r^{(k)}\rangle|^2}{\|A_i\|^2} = \max_{1\leqslant i_k\leqslant n}\left(\frac{|\langle A_i, r^{(k)}\rangle|^2}{\|A_{i_k}\|^2}\right),$$

then $i \in U_k$ for $k \in \{0, 1, 2, \cdots\}$. So U_k defined in (2.5.9) is nonempty for all iteration index k.

Proof.

$$\max_{1 \leqslant i_k \leqslant n} \frac{|\langle A_{i_k}, r^{(k)} \rangle|^2}{\|A_{i_k}\|_2^2} \geqslant \sum_{i_k=1}^n \frac{\|A_{i_k}\|^2}{\|A\|_F^2} \cdot \frac{|\langle A_{i_k}, r^{(k)} \rangle|^2}{\|A_{i_k}\|^2} = \frac{\|A^T r^{(k)}\|^2}{\|A\|_F^2},$$

and

$$\frac{|\langle A_i, r^{(k)} \rangle|^2}{\|A_i\|^2} = \max_{1 \leqslant i_k \leqslant n} \left(\frac{|\langle A_{i_k}, r^{(k)} \rangle|^2}{\|A_{i_k}\|^2} \right)$$

$$\geqslant \frac{1}{2} \max_{1 \leqslant i_k \leqslant n} \left(\frac{|\langle A_{i_k}, r^{(k)} \rangle|^2}{\|A_{i_k}\|^2} \right) + \frac{1}{2} \frac{\|A^T r^{(k)}\|^2}{\|A\|_F^2}$$

$$= \epsilon_k \|A^T r^{(k)}\|^2,$$

which implies $i \in \mathcal{U}_k$ for all iteration index k. □

Lemma 2.5.1 shows the greedy randomized Gauss-Seidel method is well defined(see Algorithm 2.5.2).

For the convergence property of the greedy randomized Gauss-Seidel method, we can establish the following theorem.

Theorem 2.5.2. *Consider the linear system (1.1.1), where the coefficient matrix $A \in \mathcal{R}^{m \times n}$ has full column rank. Then the iteration sequence $\{x^{(k)}\}_{k=0}^{\infty}$, generated by GRGS starting from any initial guess $x^{(0)} \in \mathcal{R}^n$, converges to the unique least square solution $x^* = A^+ b$ in expectation. Moreover, the solution error in expectation for the iteration sequence $\{x^{(k)}\}_{k=0}^{\infty}$ obeys*

$$E\|Ax^{(1)} - Ax^*\|^2 \leqslant \left(1 - \frac{\lambda_{\min}(A^T A)}{\|A\|_F^2} \right) \|Ax^{(0)} - Ax^*\|^2, \qquad (2.5.10)$$

and

$$E_k \|Ax^{(k+1)} - Ax^*\|_2^2$$

$$\leqslant \left[1 - \frac{1}{2}(\frac{1}{\gamma}\|A\|_F^2 + 1)\frac{\lambda_{\min}(A^T A)}{\|A\|_F^2} \right] \|Ax^{(0)} - Ax^*\|^2, k = 1, 2, 3, \cdots,$$

$$(2.5.11)$$

Algorithm 2.5.2 Greedy Randomized Gauss-Seidel (GRGS) Method

---

procedure $(A, b, x^{(0)}, K)$

    for $k = 1, 2, \cdots, K$ **do**

        Compute $\epsilon_k = \frac{1}{2} \left(\frac{1}{\|A^T r^{(k)}\|_2^2} \max\limits_{1 \leqslant i_k \leqslant n} \frac{|\langle A_{i_k}, r^{(k)} \rangle|^2}{\|A_{i_k}\|^2} + \frac{1}{\|A\|_F^2} \right)$

        Find the index set of positive integers

$$\mathcal{U}_k = \left\{ i_k \big| |\langle A_{i_k}, r^{(k)} \rangle|^2 \geqslant \epsilon_k \|A^T r^{(k)}\|^2 \|A_{i_k}\|^2 \right\}$$

        Select $i_k \in \mathcal{U}_k$ with probability $\text{Pr}(\text{Col}=i_k) = \dfrac{|\langle A_{i_k}, r^{(k)} \rangle|^2}{\sum\limits_{i_k \in \mathcal{U}_k} |\langle A_{i_k}, r^{(k)} \rangle|^2}$

        Compute α_k according to $\alpha_k = \dfrac{\langle A_{i_k}, r^{(k)} \rangle}{\|A_{i_k}\|^2}$

        Update $x^{(k+1)} = x^{(k)} + \alpha_k e_{i_k}$ and $r^{(k+1)} = r^{(k)} - \alpha_k A_{i_k}$

    end for

    Output: $x^{(K+1)}$

end procedure

---

where $\gamma = \max\limits_{1 \leqslant i \leqslant n} \sum\limits_{j=1, j \neq i}^{n} \|A_j\|_2^2$. *As a result, it holds that*

$E\|Ax^{(k)} - Ax^*\|_2^2$

$$\leqslant \left[1 - \frac{1}{2}(\frac{1}{\gamma}\|A\|_F^2 + 1) \frac{\lambda_{\min}(A^T A)}{\|A\|_F^2} \right]^{k-1} \left(1 - \frac{\lambda_{\min}(A^T A)}{\|A\|_F^2} \right) \|Ax^{(0)} - Ax^*\|_2^2.$$

$$(2.5.12)$$

Proof. Let x^* be the least square solution of $Ax = b$, which implies $A^T Ax^* = A^T b$. From the definition of the greedy randomized Gauss-Seidel method, for $k = 1, 2, \cdots$, we have

$$
\begin{aligned}
\langle A_{i_k}, r^{(k)} \rangle &= \langle A_{i_{k-1}}, b - Ax^{(k)} \rangle \\
&= \langle A_{i_{k-1}}, b \rangle - \left\langle A_{i_{k-1}}, A \left(x^{(k-1)} + \frac{\langle A_{i_{k-1}}, r^{(k-1)} \rangle}{\|A_{i_{k-1}}\|^2} e_{i_{k-1}} \right) \right\rangle \\
&= \langle A_{i_{k-1}}, b - Ax^{k-1} \rangle - \langle A_{i_{k-1}}, r^{(k-1)} \rangle \\
&= 0.
\end{aligned}
$$

It then follows for $k = 1, 2, \cdots$ that

$$
\begin{aligned}
\epsilon_k \|A\|_F^2 &= \frac{1}{2} \frac{\|A\|_F^2}{\|A^T r^{(k)}\|^2} \max_{1 \leqslant i_k \leqslant n} \left(\frac{|\langle A_{i_k}, r^{(k)} \rangle|^2}{\|A_{i_k}\|^2} \right) + \frac{1}{2} \\
&= \frac{\displaystyle\max_{1 \leqslant i_k \leqslant n} \left(\frac{|\langle A_{i_k}, r^{(k)} \rangle|^2}{\|A_{i_k}\|^2} \right)}{\displaystyle 2 \sum_{\substack{i_k=1 \\ i_k \neq i_{k-1}}}^{n} \frac{\|A_{i_k}\|^2}{\|A\|_F^2} \cdot \frac{|\langle A_{i_k}, r^{(k)} \rangle|^2}{\|A_{i_k}\|^2}} + \frac{1}{2} \\
&= \frac{\displaystyle\max_{1 \leqslant i_k \leqslant n} \left(\frac{|\langle A_{i_k}, r^{(k)} \rangle|^2}{\|A_{i_k}\|^2} \right)}{\displaystyle 2 \sum_{i_k=1}^{n} \frac{\|A_{i_k}\|^2}{\|A\|_F^2} \cdot \frac{|\langle A_{i_k}, r^{(k)} \rangle|^2}{\|A_{i_k}\|^2}} + \frac{1}{2} \\
&\geqslant \frac{1}{2} \left(\frac{\|A\|_F^2}{\displaystyle\sum_{\substack{i_k=1 \\ i_k \neq i_{k-1}}}^{n} \|A_{i_k}\|^2} + 1 \right) \\
&\geqslant \frac{1}{2} \left(\frac{1}{\gamma} \|A\|_F^2 + 1 \right),
\end{aligned}
$$

where $\displaystyle\gamma = \max_{1 \leqslant i_{k-1} \leqslant n} \sum_{\substack{i_k=1 \\ i_k \neq i_{k-1}}}^{n} \|A_{(i_k)}\|^2 < \|A\|_F^2$.

For $k = 0$, we can get

$$\epsilon_0 \|A\|_F^2 = \frac{\max\limits_{1 \leqslant i_0 \leqslant n} \left(\frac{|\langle A_{i_0}, r^{(k)} \rangle|^2}{\|A_{i_0}\|^2} \right)}{2 \sum\limits_{i_0=1}^{n} \frac{\|A_{i_0}\|^2}{\|A\|_F^2} \cdot \frac{|\langle A_{i_0}, r^{(k)} \rangle|^2}{\|A_{i_k}\|^2}} + \frac{1}{2}$$

$$\geqslant \frac{1}{2} + \frac{1}{2}$$

$$= 1.$$

Again, from the definition of the greedy Gauss-Seidel method we obtain

$$Ax^{(k+1)} - Ax^{(k)} = \frac{\langle A_{i_k}, r^{(k)} \rangle}{\|A_{i_k}\|^2} A_{i_k},$$

which implies that $Ax^{(k+1)} - Ax^{(k)}$ is parallel to A_{i_k}. Since

$$A_{i_k}^T Ax^* = A_{i_k}^T b \tag{2.5.13}$$

it holds that

$$A_{i_k}^T (Ax^{(k+1)} - Ax^*) = A_{i_k}^T A \left(x^{(k)} + \frac{\langle A_{i_k}, r^{(k)} \rangle}{\|A_{i_k}\|^2} e_{i_k} \right) - A_{i_k}^T Ax^*$$

$$= A_{i_k}^T Ax^{(k)} + \langle A_{i_k}, r^{(k)} \rangle - A_{i_k}^T b$$

$$= A_{i_k}^T (Ax^{(k)} - b) - \langle A_{i_k}, r^{(k)} \rangle$$

$$= 0,$$

which shows that $Ax^{(k+1)} - Ax^*$ is orthogonal to A_{i_k}. Therefore, we know that $Ax^{(k+1)} - Ax^*$ is perpendicular to $Ax^{(k+1)} - Ax^{(k)}$ for any $k \geqslant 0$ or, in other words, $(Ax^{(k+1)} - Ax^*)^T (Ax^{(k+1)} - Ax^{(k)}) = 0$. It follows that

$$\|Ax^{(k+1)} - Ax^*\|^2 = \|Ax^{(k)} - Ax^*\|^2 - \|Ax^{(k+1)} - Ax^{(k)}\|^2.$$

Based on this equality, we have

$$E_k \|Ax^{(k+1)} - Ax^*\|^2$$

$$= \|Ax^{(k)} - Ax^*\|^2 - E_k \|Ax^{(k+1)} - Ax^{(k)}\|^2$$

$$= \|Ax^{(k)} - Ax^*\|^2 - \sum_{i_k \in U_k} \frac{|\langle A_{i_k}, r^{(k)} \rangle|^2}{\sum_{i_k \in U_k} |\langle A_{i_k}, r^{(k)} \rangle|^2} \cdot \frac{\|A_{i_k}^T A(x^{(k)} - x^*)\|^2}{\|A_{i_k}\|^2}$$

$$\leqslant \|Ax^{(k)} - Ax^*\|^2 - \epsilon_k \|A^T b - A^T Ax^{(k)}\|^2.$$

Here the last inequality is achieved with the use of the definitions in (2.5.8) and (2.5.9) of the quantity ϵ_k and the set $\mathcal{U}_k$, which leads to the inequality

$$\|A_{i_k}^T A(x^{(k)} - x^*)\|^2 \geqslant \epsilon_k \|A^T r^{(k)}\|^2 \|A_{i_k}\|^2, \quad \forall i_k \in \mathcal{U}_k.$$

We can then obtain an estimate as follows:

$$\begin{aligned}
E_k \|Ax^{(k+1)} - Ax^*\|^2 &\leqslant \|Ax^{(k)} - Ax^*\|^2 - \epsilon_k \|A^T b - A^T Ax^{(k)}\|^2 \\
&= \|Ax^{(k)} - Ax^*\|^2 - \epsilon_k \|A^T A(x^* - x^{(k)})\|^2 \\
&\leqslant (1 - \epsilon_k \lambda_{\min}(A^T A))\|Ax^{(k)} - Ax^*\|^2.
\end{aligned}$$

Here in the last inequality we have used the estimate

$$\|Au\|^2 \geqslant \lambda_{\min}(A^T A))\|u\|^2, \text{ and } \lambda(A^T A) = \lambda(AA^T),$$

which holds true for any $u \in \mathcal{R}^n$ belonging to the column space of A^T. By making use of

$$\epsilon_0 \geqslant \frac{1}{\|A\|_F^2}$$

and

$$\epsilon_k \geqslant \frac{1}{2}\left(\frac{1}{\gamma}\|A\|_F^2 + 1\right)\frac{1}{\|A\|_F^2}, \quad k = 1, 2, \cdots,$$

we can obtain the estimate (2.5.10) and (2.5.11) about the convergence rate of the greedy randomized Gauss-Seidel iteration sequence.

Finally, by taking the full expectation on both sides of (2.5.11), we have

$$E\|Ax^{(k+1)} - Ax^*\|^2 \leqslant \left[1 - \frac{1}{2}\left(\frac{1}{\gamma}\|A\|_F^2 + 1\right)\frac{\lambda_{\min}(A^T A)}{\|A\|_F^2}\right]E\|Ax^{(k)} - Ax^*\|^2,$$

for $k = 1, 2, \cdots$. By induction on the iteration index k, we straightforwardly obtain the estimate (2.5.12). $\square$

Remark 2.5.2. As $\frac{1}{2}\left(\frac{1}{\gamma}\|A\|_F^2 + 1\right) \geqslant 1$, it holds that

$$1 - \frac{1}{2}\left(\frac{1}{\gamma}\|A\|_F^2 + 1\right)\frac{\lambda_{\min}(A^T A)}{\|A\|_F^2} < 1 - \frac{\lambda_{\min}(A^T A)}{\|A\|_F^2}.$$

Hence, the convergence factor of the greedy randomized Gauss-Seidel method is smaller and faster than randomized Gauss-Seidel method.

Remark 2.5.3. *If the system $Ax = b$ is consistent, so $Ax^* = b$. In this setting, both greedy randomized Kaczmarz (GRK) method and greedy randomized Gauss-Seidel (GRGS) methods are essentially eqivalent (without computational considerations). If $Ax = b$ is inconsistent, so $Ax^* + r = b$, where r is such that $A^T r = 0$. In this setting, GRK does not converge to the least square solution x^*, but GRGS does at the same linear rate.*

2.6   REK and REGS

Needell[47] later studied the inconsistent system of linear equation $Ax = b = \bar{b} + \delta$, where δ is arbitrary error vector, and showed that RK does not converge in expectation to the least squares solution for this inconsistent systems, but reaches an error threshold dependent on the matrix A with the same rate as in the error-free case ($\delta = 0$). That is,

$$E\|x^{(k)} - x^*\| \leqslant \left(1 - \frac{1}{\|A^{-1}\|^2 \|A\|_F^2}\right)^{k/2} \|x^{(0)} - x^*\| + \max_i \frac{\delta_i \|A^{-1}\| \|A\|_F}{\|a_i\|}.$$

In 2013, Zouzias and Freris[40] gave two modifications: the full column rank assumption on the input matrix is dropped (the full column rank assumption is not necessary) and the additive term that is related to δ_i is improved to $\|\delta\|^2 / \|A\|_F^2$. That is,

$$E\|x^{(k)} - x^*\|^2 \leqslant \left(1 - \frac{1}{\kappa_F^2(A)}\right)^k \|x^{(0)} - x^*\|^2 + \frac{\|\delta\|^2}{\sigma_{min}^2},$$

where $\kappa_F(A) = \|A\|_F \|A^+\|_2$ and σ_{min} is the nonzero singular value of A.

To remedy the non-convergence of the RK to the inconsistent system of linear equation, Zouzias and Freris[40] proposed the **Randomized Extended Kaczmarz (REK)** algorithm to solve linear systems in all settings.

Given a least squares problem, from the inequality with $\delta = b_{\mathcal{R}(A)^\perp}$ (and $\bar{b} = b_{\mathcal{R}(A)}$) we can know that the randomized Kaczmarz algorithm works well for least squares problems whose least squares error is very close to zero, i.e., $\delta \approx 0$. Roughly speaking, in this case the randomized Kaczmarz algorithm approaches the minimum l_2-norm least squares solution up to an additive error that depends on the distance between b and the column space of A.

The main observation is that it is possible to efficiently reduce the norm of the noisy part of b, $\delta = b_{\mathcal{R}(A)^\perp}$, using an orthogonal projection algorithm (see the following algorithm), and then apply the randomized Kaczmarz algorithm on a new linear system whose right-hand side vector is now arbitrarily close to the column space of A, i.e., $Ax \approx b_{\mathcal{R}(A)}$. This idea together with the observation that the least squares solution of the latter linear system is equal (in the limit) to the least squares solution of the original system implies a randomized algorithm for solving least squares.

The randomized extended Kaczmarz algorithm is presented which is a specific combination of the randomized orthogonal projection algorithm together with the randomized Kaczmarz algorithm.

Algorithm 2.6.1 Randomized Exteneded Kaczmarz (REK) Algorithm

---

> **procedure** $(A,\ b,\ x^{(0)},\ K,\ \epsilon > 0)$
>
>   Set $x^{(0)} = 0$ and $z^{(0)} = b$
>
>   **for** $k = 0, 1, \cdots, K-1$ **do**
>
>   Select $i_k \in [m]$ with probability $\Pr(\text{row}=i_k) = \frac{\|a_{i_k}\|^2}{\|A\|_F^2}$
>
>   Select $j_k \in [n]$ with probability $\Pr(\text{colomn}=j_k) = \frac{\|A_{j_k}\|^2}{\|A\|_F^2}$
>
>   Set $z^{(k+1)} = z^{(k)} - \frac{\langle A_{j_k}, z^{(k)} \rangle}{\|A_{j_k}\|_2^2} A_{j_k}$
>
>   Set $x^{(k+1)} = x^{(k)} + \frac{b_{i_k} - z_{i_k}^{(k)} - \langle a_{i_k}, x^{(k)} \rangle}{\|a_{i_k}\|^2} a_{i_k}$
>
>   Check every $8 \min(m, n)$ iterations and terminate if it holds:
>
> $$\frac{\|Ax^{(k)} - (b - z^{(k)})\|}{\|A\|_F^2 \|x^{(k)}\|} \leqslant \epsilon \quad and \quad \frac{\|A^T z^{(k)}\|}{\|A\|_F^2 \|x^{(k)}\|} \leqslant \epsilon.$$
>
>   **end for**
>
>   Output: $x^{(K)}$
>
> **end procedure**

The algorithm converges in mean square to the minimum l_2-norm solution vector x_{LS}. The proposed algorithm consists of two components. The first component consisting of steps 5 and 6 is responsible for implicitly maintaining an approximation to $b_{\mathcal{R}(A)}$ formed by $b - z^{(k)}$. The second component, consisting of steps 4 and 7, applies the randomized Kaczmarz algorithm with input A and the current approximation $b - z^{(k)}$ of $b_{\mathcal{R}(A)}$, i.e., applies one step of the randomized Kaczmarz on the system $Ax = b - z^{(k)}$. Since $b - z^{(k)}$ converges to $b_{\mathcal{R}(A)}$, $x^{(k)}$ will eventually converge to the minimum Euclidean norm solution of $Ax = b_{\mathcal{R}(A)}$ which equals $x_{LS} = A^+ b$.

For **REK**, we have the following theorem of rate of convergence.

Theorem 2.6.1. *After $K > 1$ iterations, in exact arithmetic, REK algorithm with input A (possibly rank deficient) and b computes a vector $x^{(K)}$ such that*

$$E\|x^{(K)} - x_{LS}\|^2 \leqslant \left(1 - \frac{\sigma_{\min}^2(A)}{\|A\|_F^2}\right)^{\lfloor K/2 \rfloor} \left(1 + 2\frac{\sigma_{\min}^2(A)}{\sigma_{\max}^2(A)}\|x_{LS}\|^2\right).$$

REGS. When $m < n$, there are fewer constraints than variables, and the consistent system $Ax = b$ has infinitely many solutions. In this case, especially if we have no prior reason to believe any additional sparsity in the signal structure, we are often interested in finding the least Euclidean norm solution

$$x_{LN} = arg \min_x \|x\|_2 \quad s.t. \quad Ax = b. \tag{2.6.1}$$

While **RGS** converges to x_{LN} in the overcomplete setting, the undercomplete setting does not converge to x_{LN}. We have known that **RK** does converge to x_{LN} without any extensions in this setting.

In 2015, Ma & Needell etc.[46] constructed an extension to **RGS** that parallels **REK**, which unlike **RGS** does converge to x_{LN} (just as **REK**, unlike **RK**, which converges to x_{LS}). Some desired properties for this algorithm include that it should also converge linearly, without requiring much extra computation, and working well in simulations. A summary of this unified theory is provided in the following Table 2.1

Table 2.1: Summary of convergence properties

Method	$rank(A) = n,$ consistent $\Rightarrow m \geqslant n,$ unique x_{uniq}	$rank(A) < n,$ consistent $\Rightarrow m \gtreqless n,$ unique x_{LN}	$rank(A) = n,$ inconsistent $\Rightarrow m \geqslant n,$ unique x_{LS}	$rank(A) < n,$ inconsistent $\Rightarrow m \gtreqless n,$ unique x_{LNLS}
RK	Yes	Yes	No	No
REK	Yes	Yes	Yes	Yes
RGS	Yes	No	Yes	No
REGS	Yes	Yes	Yes	Yes

The **Randomized Extended Gauss-Seidel (REGS)** resolves rank deficient consistent problem, much as REK did for RK in the case of full rank inconsistent systems. The method chooses a random row and column of A exactly as in REK, and then updates at every iteration

$$w^{(k+1)} = \frac{\langle A_j, b - Ax^{(k)} \rangle}{\|A_j\|^2} e_j,$$
$$x^{(k+1)} = x^{(k)} + w^{(k+1)},$$
$$P_i = I_n - \frac{a_i a_i^T}{\|a_i\|^2},$$
$$z^{(k+1)} = P_i(z^{(k)} + w^{(k+1)}),$$
$$x_{LN}^{(k+1)} = x^{(k+1)} - z^{(k+1)}.$$

Here, I_n denotes the $n \times n$ identity matrix. This extension works for all variations of linear systems and was proven to converge linearly in expectation by Ma & Needell etc.[46].

In 2019, K. Du[48] presented tight upper bounds (in the sense that there is a linear system $Ax = b$ for which the inequality for upper bound holds with equality) for the convergence of REK and REGS. These bounds hold for all types of linear systems (consistent or inconsistent, overdetermined or underdetermined, A has full column rank or not) and are better than the existing ones.

Remark 2.6.1. *A summary of convergence properties for the rank full and* *consistent setting, rank full and inconsistent setting, rank deficient and consistent settings, and rank deficient and inconsistent settings is as follows. In Table 2.1, we write x_{uniq} to denote the solution to (1.1.1) in the full rank consistent setting, with x_{LS} and x_{LN} being defined as the ordinary (unique) least squares solution and the (unique) least Euclidean norm solution. When x_{LNLS} is defined as the least norm least squares solution, **REK and REGS still converge to** x_{LNLS}.*

Remark 2.6.2. *How to apply RGS (REK, REGS) to sparse solution?*

Remark 2.6.3. *In[49], authors analyzed the convergence rates of variants of RK and RGS for ridge regression, which corresponds to solving the convex optimization problem:* $\min\limits_{x \in \mathcal{R}^n} \|b - Ax\|_2^2 + \mu\|x\|_2^2$, *showing linear convergence in expectation, and they showed that when $m > n$, it should randomize over columns (RGS), and when $m < n$, it should be randomized over rows (RK). How about to problem* $\min\limits_{x \in \mathcal{R}^n} \|b - Ax\|_2^2 + \mu\|x\|_1$? *Especially, when* $rank(A) < n$ *and the system is inconsistent, how to get the* **sparsest least square solution***?*

2.7 RK and RGS with Oblique Projection

2.7.1 Random Kaczmarz Method with Oblique Projection

If the row in Algorithm 1.4.1 is randomly selected, we get a randomized Kaczmarz method with oblique projection (RKO) [31].

Lemma 2.7.1. *Consider the linear system (1.1.1)($m > 2$). Let $x^{(0)} \in \mathcal{R}^n$ be an arbitrary initial approximation, $\tilde{x}$ is a solution of the system (1.1.1), which satisfies $P_{\mathcal{N}(A)}(\tilde{x}) = P_{\mathcal{N}(A)}(x^{(0)})$, then we obtain the bound on the following expected conditional on the first k ($k \geqslant 2$) interation of the RKO as follows*

(a) If $Pr(a_{i_k}) = \frac{1}{m-2}$, then $E_k \frac{|r_{i_k}^{(k)}|^2}{\|w_{i_k}\|^2} \geqslant \frac{\sigma_{min}^2(A)\|x^{(k)} - \tilde{x}\|^2}{(m-2)(\|A\|_F^2 - \sigma_{min}^2(A))}$.

(b) If $Pr(a_{i_k}) = \frac{|r_{i_k}^{(k)}|^2}{\|r^{(k)}\|^2}$, then $E_k \frac{|r_{i_k}^{(k)}|^2}{\|w_{i_k}\|^2} \geqslant \frac{\sigma_{min}^2(A)\|x^{(k)} - \tilde{x}\|^2}{\|A\|_F^2 - \sigma_{min}^2(A)}$.

Proof. (a) If $Pr(a_{i_k}) = \frac{1}{m-2}$,

$$
\begin{aligned}
E_k \frac{|r_{i_k}^{(k)}|^2}{\|w_{i_k}\|^2} &\overset{(ii)}{=} \frac{1}{m-2} \sum_{\substack{s=1 \\ s \neq i_{k-2}, i_{k-1}}}^{m} \frac{r_s^2}{\|w_s\|^2} \\
&\overset{(iii)}{\geqslant} \frac{1}{m-2} \frac{\sum_{\substack{s=1 \\ s \neq i_{k-2}, i_{k-1}}}^{m} r_s^2}{\sum_{\substack{s=1 \\ s \neq i_{k-2}, i_{k-1}}}^{m} \|w_s\|^2} \\
&\geqslant \frac{1}{m-2} \frac{\sum_{s=1}^{m} r_s^2}{\sum_{s=1}^{m} \|a_s\|^2 - \frac{\langle a_s, a_{i_k} \rangle^2}{\|a_{i_k}\|^2}} \\
&= \frac{1}{m-2} \frac{\|b - Ax^{(k)}\|^2}{\|A\|_F^2 - \frac{\|Aa_{i_k}\|^2}{\|a_{i_k}\|^2}} \\
&\overset{(iv)}{\geqslant} \frac{\sigma_{min}^2(A)\|x^{(k)} - \tilde{x}\|^2}{(m-2)\left(\|A\|_F^2 - \sigma_{min}^2(A)\right)},
\end{aligned}
\tag{2.7.1}
$$

where the equality (ii) holds as $r_{i_{k-2}} = r_{i_{k-1}} = 0$, and the inequality (iii) uses the conclusion of $\frac{|b_1|}{|a_1|} + \frac{|b_2|}{|a_2|} \geqslant \frac{|b_1|+|b_2|}{|a_1|+|a_2|}$ (if $|a_1| > 0$, $|a_2| > 0$).

From the iteration of RKO, i.e. step 3, step 5 and step 11 of Algorithm 2.7.1, it can be obtained that $P_{\mathcal{N}(A)}(x^{(k)}) = P_{\mathcal{N}(A)}(x^{(k-1)}) = \cdots = P_{\mathcal{N}(A)}(x^{(0)})$, and by the fact that $P_{\mathcal{N}(A)}(x^{(0)}) = P_{\mathcal{N}(A)}(\tilde{x})$, we can deduce that $x^{(k)} - \tilde{x} \in \mathcal{R}(A^T)$. The inequality (iv) uses the conclusion of $\|Az\| \geqslant \sigma_{min}(A)\|z\|$, if $z \in \mathcal{R}(A^T)$.

(b) If $Pr(a_{i_k}) = \frac{|r_{i_k}^{(k)}|^2}{\|r^{(k)}\|^2}$,

$$
\begin{aligned}
E_k \frac{|r_{i_k}^{(k)}|^2}{\|w_{i_k}\|^2} &= \sum_{\substack{s=1 \\ s \neq i_{k-2}, i_{k-1}}}^{m} \frac{|r_s^{(k)}|^2}{\|r^{(k)}\|^2} \frac{|r_s^{(k)}|^2}{\|w_s\|^2} \\
&\overset{(v)}{\geqslant} \frac{1}{\|r^{(k)}\|^2} \frac{\left(\sum_{\substack{s=1 \\ s \neq i_{k-2}, i_{k-1}}}^{m} |r_s^{(k)}|^2\right)^2}{\sum_{\substack{s=1 \\ s \neq i_{k-2}, i_{k-1}}}^{m} \|w_s\|^2} \\
&= \frac{\|r^{(k)}\|^2}{\sum_{s=1}^{m} \left(\|a_s\|^2 - \frac{\langle a_s, a_{i_k}\rangle^2}{\|a_{i_k}\|^2}\right) - \left(\|a_{i_{k-1}}\|^2 - \frac{\langle a_{i_{k-1}}, a_{i_k}\rangle^2}{\|a_{i_k}\|^2}\right)}
\end{aligned}
$$

$$
\begin{aligned}
&= \frac{\|r^{(k)}\|^2}{\sum\limits_{s=1}^{m}\left(\|a_s\|^2-\frac{\langle a_s,a_{i_k}\rangle^2}{\|a_{i_k}\|^2}\right)-\|a_{i_{k-1}}\|^2 sin^2\theta} \\
&\geqslant \frac{\|r^{(k)}\|^2}{\sum\limits_{s=1}^{m}\left(\|a_s\|^2-\frac{\langle a_s,a_{i_k}\rangle^2}{\|a_{i_k}\|^2}\right)} \\
&= \frac{\|b-Ax^{(k)}\|^2}{\|A\|_F^2-\frac{\|Aa_{i_k}\|^2}{\|a_{i_k}\|^2}} \\
&\geqslant \frac{\sigma_{min}^2(A)\|x^{(k)}-\tilde{x}\|^2}{\|A\|_F^2-\sigma_{min}^2(A)},
\end{aligned}
\tag{2.7.2}
$$

where inequality (v) uses the conclusion of

$$
\sum_{i=1}^{m}\frac{x_i^{\eta+1}}{y_i^{\eta}} \geqslant \frac{\left(\sum\limits_{i=1}^{m}x_i\right)^{\eta+1}}{\left(\sum\limits_{i=1}^{m}y_i\right)^{\eta}}.
\tag{2.7.3}
$$

for $m \in \mathbb{N}^+, \eta > 0, x_i \geqslant 0, y_i > 0, i \in [m]$. $\qquad\square$

Theorem 2.7.1. *Consider the linear system (1.1.1)($m > 2$). Let $x^{(0)} \in \mathcal{R}^n$ be an arbitrary initial approximation, $\tilde{x}$ is a solution of the system (1.1.1), which satisfies $P_{N(A)}(\tilde{x}) = P_{N(A)}(x^{(0)})$. Then the iteration sequence $\{x^{(k)}\}_{k=2}^{\infty}$ generated by the RKO method obeys*
(a) If $Pr(a_{i_k}) = \frac{1}{m-2}$,

$$
E_k\|x^{(k+1)} - \tilde{x}\|^2 \leqslant \left(1 - \frac{\sigma_{min}^2(A)}{(m-2)(\|A\|_F^2 - \sigma_{min}^2(A))}\right)\|x^{(k)} - \tilde{x}\|^2 \ (k \geqslant 2).
$$

(b) If $Pr(a_{i_k}) = \frac{|r_{i_k}^{(k)}|^2}{\|r^{(k)}\|^2}$,

$$
E_k\|x^{(k+1)} - \tilde{x}\|^2 \leqslant \left(1 - \frac{\sigma_{min}^2(A)}{\|A\|_F^2 - \sigma_{min}^2(A)}\right)\|x^{(k)} - \tilde{x}\|^2 \ (k \geqslant 2).
$$

Proof. With the use of the orthogonality, the following equation holds:

$$
\begin{aligned}
\|x^{(k+1)} - \tilde{x}\|^2 &= \|x^{(k)} - \tilde{x}\|^2 - \|x^{(k+1)} - x^{(k)}\|^2 \\
&= \|x^{(k)} - \tilde{x}\|^2 - \|\alpha_{i_k}^{(k)} w_{i_k}\|^2 \\
&= \|x^{(k)} - \tilde{x}\|^2 - \left\|\frac{r_{i_k}^{(k)}}{h_{i_k}} w_{i_k}\right\|^2 \\
&\overset{(v)}{=} \|x^{(k)} - \tilde{x}\|^2 - \frac{|r_{i_k}^{(k)}|^2}{\|w_{i_k}\|^2},
\end{aligned}
\tag{2.7.4}
$$

Algorithm 2.7.1 RK with Oblique Projection (RKO)

---

procedure $(A,\ b,\ x^{(0)},\ K,\ \varepsilon > 0,\ M(i) = \|a_i\|^2,\ i \in [m])$

    Randomly select i_0, $i_1 \neq i_0$, compute $x^{(1)} = x^{(0)} + \frac{b_{i_0} - \langle a_{i_0}, x^{(0)} \rangle}{M(i_0)} a_{i_0}$ and

    compute $x^{(2)} = x^{(1)} + \frac{b_{i_1} - \langle a_{i_1}, x^{(1)} \rangle}{\|a_{i_1}\|^2 - \frac{\langle a_{i_0}, a_{i_1} \rangle^2}{\|a_{i_0}\|^2}} \left(a_{i_1} - \frac{\langle a_{i_0}, a_{i_1} \rangle}{\|a_{i_0}\|^2} a_{i_0} \right)$

    for $k = 2, 3, \cdots, K$ **do**

        Randomly select i_k $(i_k \neq i_{k-1}, i_{k-2})$

        Compute $D_{i_k} = \langle a_{i_k}, a_{i_{k-1}} \rangle$, $r_{i_k}^{(k)} = b_{i_k} - \langle a_{i_k}, x^{(k)} \rangle$

        Compute $w_{i_k} = a_{i_k} - \frac{D_{i_k}}{M(i_{k-1})} a_{i_{k-1}}$, $h_{i_k} = M(i_k) - \frac{D_{i_{k-1}}^2}{M(i_{k-1})}$ $(=\|w_{i_k}\|^2)$

        if $h_{i_k} > \varepsilon$ **then**

            $\alpha_{i_k}^{(k)} = \frac{r_{i_k}^{(k)}}{h_{i_k}}$ and $x^{(k+1)} = x^{(k)} + \alpha_{i_k}^{(k)} w_{i_k}$

        end if

    end for

    Output $x^{(K+1)}$

end procedure

---

where the equality (vi) holds as $h_{i_k} = \|w_{i_k}\|^2$. Making conditional expectation on both sides, and applying Lemma 2.7.1, we get

(a) If $Pr(a_{i_k}) = \frac{1}{m-2}$,

$$
\begin{aligned}
E_k \|x^{(k+1)} - \tilde{x}\|^2 &= \|x^{(k)} - \tilde{x}\|^2 - E_k \left[\frac{(r_{i_{k+1}}^{(k)})^2}{\|w_{i_k}\|^2} \right] \\
&\leqslant \|x^{(k)} - \tilde{x}\|^2 - \frac{\sigma_{min}^2(A)\|x^{(k)} - \tilde{x}\|^2}{(m-2)(\|A\|_F^2 - \sigma_{min}^2(A))} \\
&= (1 - \frac{\sigma_{min}^2(A)}{(m-2)(\|A\|_F^2 - \sigma_{min}^2(A))})\|x^{(k)} - \tilde{x}\|.
\end{aligned}
$$

(b) If $Pr(a_{i_k}) = \frac{|r_{i_k}^{(k)}|^2}{\|r^{(k)}\|^2}$,

$$
E_k \|x^{(k+1)} - \tilde{x}\|^2 = \|x^{(k)} - \tilde{x}\|^2 - E_k \left[\frac{(r_{i_{k+1}}^{(k)})^2}{\|w_{i_k}\|^2} \right]
$$

$$
\leqslant \|x^{(k)} - \tilde{x}\|^2 - \frac{\sigma_{min}^2(A)\|x^{(k)} - \tilde{x}\|^2}{\|A\|_F^2 - \sigma_{min}^2(A)} = (1 - \frac{\sigma_{min}^2(A)}{\|A\|_F^2 - \sigma_{min}^2(A)})\|x^{(k)} - \tilde{x}\|.
$$

It yields the desired result. □

Remark 2.7.1. *In particular, after unitizing the rows of matrix A (this does not lose the generality of the system), we can get from Lemma 2.7.1:*

$$
\begin{aligned}
E_k \frac{|r_{i_k}^{(k)}|^2}{\|w_{i_k}\|^2} &= \frac{1}{m-2} \sum_{\substack{s=1 \\ s \neq i_{k-2}, i_{k-1}}}^{m} \frac{(r_s^{(k)})^2}{\|a_s\|^2 - \frac{\langle a_s, a_{i_k} \rangle^2}{\|a_{i_k}\|^2}} \\
&= \sum_{\substack{s=1 \\ s \neq i_{k-2}, i_{k-1}}}^{m} \frac{\|a_s\|^2}{\|A\|_F^2 - 2} \frac{(r_s^{(k)})^2}{\|a_s\|^2 - \frac{\langle a_s, a_{i_k} \rangle^2}{\|a_{i_k}\|^2}} \\
&\geqslant \sum_{\substack{s=1 \\ s \neq i_{k-2}, i_{k-1}}}^{m} \frac{1}{\|A\|_F^2 - 2} \frac{(r_s^{(k)})^2}{1 - \gamma_{i_k}^2} \\
&\geqslant \frac{\sigma_{min}^2(A)\|x^{(k)} - \tilde{x}\|_2^2}{(1 - \gamma_{i_k}^2)(\|A\|_F^2 - 2)}, k = 2, 3, \dots
\end{aligned}
$$

where $\gamma_{i_k} = \min\limits_{s \neq i_{k-2}, i_{k-1}} |\langle a_s, a_{i_k} \rangle|$. Then we get from Theorem 2.7.1:

$$E_k \|x^{(k+1)} - \tilde{x}\|^2 \leqslant \left(1 - \frac{\sigma_{min}^2(A)}{(1 - \gamma_{i_k}^2)(\|A\|_F^2 - 2)}\right) \|x^{(k)} - \tilde{x}\|^2.$$

The convergence of the RK method in [39] meets:

$$E_k \|x^{(k+1)} - \tilde{x}\|^2 \leqslant \left(1 - \frac{\sigma_{min}^2(A)}{\|A\|_F^2}\right) \|x^{(k)} - \tilde{x}\|^2.$$

Obviously, the convergence speed of the RKO method is faster than the RK method.

2.7.2 RGS with Oblique Projection

If the columns (whose residual entries are not 0) in algorithm 1.5.1 are selected uniformly and randomly, we get a randomized Gauss-Seidel method with oblique direction (RGSO) [50].

Lemma 2.7.2. *Consider the linear least-squares problem (1.5.2), where the coefficient $A \in \mathcal{R}^{m \times n}$, $b \in \mathcal{R}^m$ is a given vector, and $\tilde{x}$ is any solution to the system (1.5.2) , then we obtain the bound on the following expected conditional on the first k $(k \geqslant 2)$ iteration of the RGSO*

$$E_k \frac{(A_{i_{k+1}}^T r^{(k)})^2}{g_{i_k}} \geqslant \frac{1}{n-2} \frac{\sigma_{min}^2(A) \|\tilde{x} - x^{(k)}\|_{A^T A}^2}{\|A\|_F^2 - \sigma_{min}^2(A)}.$$

Proof. For the RGSO mthod, it is easy to get that $A_{i_k}^T r^{(k)} = 0$ $(k = 1, 2, \cdots)$ and $A_{i_{k-1}}^T r^{(k)} = 0$ $(k = 2, 3, \cdots)$ are still valid.

$$E_k \frac{(A_{i_{k+1}}^T r^{(k)})^2}{g_{i_k}} = \frac{1}{n-2} \sum_{\substack{s=1 \\ s \neq i_k, i_{k-1}}}^{n} \frac{(A_s^T r^{(k)})^2}{\|A_s\|^2 - \frac{\langle A_s, A_{i_k} \rangle^2}{\|A_{i_k}\|^2}}$$

Algorithm 2.7.2 RGS with Oblique Direction (RGSO)

---

procedure $(A, \ b, \ x^{(0)}, \ K, \ \varepsilon > 0, \ N(i) = \|A_i\|^2, \ i \in [n])$

Randomly select i_1, and compute $r^{(0)} = b - Ax^{(0)}$, $\alpha_0 = \frac{\langle A_{i_1}, r^{(0)} \rangle}{\|A_{i_1}\|^2}$,
$x^{(1)} = x^{(0)} + \alpha_0 e_{i_1}$, and randomly select $i_2 \neq i_1$, and compute $r^{(1)} = r^{(0)} - \alpha_0 A_{i_1}$, $\alpha_1 = \frac{\langle A_{i_2}, r^{(1)} \rangle}{\|A_{i_2}\|^2 - \frac{\langle A_{i_1}, A_{i_2} \rangle^2}{\|A_{i_1}\|^2}}$, and $\beta_1 = -\frac{\langle A_{i_1}, A_{i_2} \rangle}{\|A_{i_1}\|^2} \alpha_1$

Compute $x^{(2)} = x^{(1)} + \alpha_1 e_{i_2} + \beta_1 e_{i_1}$, and $r^{(2)} = r^{(1)} - \alpha_1 A_{i_2} - \beta_1 A_{i_1}$

for $k = 2, 3, \cdots, K - 1$ **do**

   Randomly select i_{k+1} $(i_{k+1} \neq i_k, i_{k-1})$

   Compute $G_{i_k} = \langle A_{i_k}, A_{i_{k+1}} \rangle$ and $g_{i_k} = N(i_{k+1}) - \frac{G_{i_k}}{N(i_k)} G_{i_k}$

   if $g_{i_k} > \varepsilon$ **then**

      Compute $\alpha_k = \frac{\langle A_{i_{k+1}}, r^{(k)} \rangle}{g_{i_k}}$, $\beta_k = -\frac{G(i_k)}{N(i_k)} \alpha_k$,
      and $x^{(k+1)} = x^{(k)} + \alpha_k e_{i_{k+1}} + \beta_k e_{i_k}$

      Compute $r^{(k+1)} = r^{(k)} - \alpha_k A_{i_{k+1}} - \beta_k A_{i_k}$

   end if

end for

Output: $x^{(K)}$

end procedure

$$
\geqslant \frac{1}{n-2} \frac{\displaystyle\sum_{s=1, s\neq i_k, i_{k-1}}^{n} (A_s^T r^{(k)})^2}{\displaystyle\sum_{\substack{s=1 \\ s\neq i_k, i_{k-1}}}^{n} (\|A_s\|^2 - \frac{\langle A_s, A_{i_k}\rangle^2}{\|A_{i_k}\|^2})} \left(= \frac{1}{n-2} \frac{\displaystyle\sum_{s=1}^{n} (A_s^T r^{(k)})^2}{\displaystyle\sum_{s=1}^{n} (\|A_s\|^2 - \frac{\langle A_s, A_{i_k}\rangle^2}{\|A_{i_k}\|^2})} \right)
$$

$$
= \frac{1}{n-2} \frac{\|A^T A(\tilde{x}-x^{(k)})\|^2}{\|A\|_F^2 - \frac{\|A^T A_{i_k}\|^2}{\|A_{i_k}\|^2}}
$$

$$
\geqslant \frac{1}{n-2} \frac{\sigma_{min}^2(A)\|\tilde{x}-x^{(k)}\|_{A^T A}^2}{\|A\|_F^2 - \sigma_{min}^2(A)}, \quad k = 2, 3, \dots
$$

The first inequality uses the conclusion of $\frac{|b_1|}{|a_1|} + \frac{|b_2|}{|a_2|} \geqslant \frac{|b_1|+|b_2|}{|a_1|+|a_2|}$ (if $|a_1| > 0$, $|a_2| > 0$), and the second one uses the conclusion of $\|A^T z\|_2^2 \geqslant \sigma_{min}^2(A)\|z\|_2^2$, if $z \in \mathcal{R}(A)$. $\qquad \square$

Theorem 2.7.2. *Consider the linear least-squares problem (1.5.2), where the coefficient $A \in \mathcal{R}^{m \times n}$, $b \in \mathcal{R}^m$ is a given vector, and $\tilde{x}$ is any least-squares solution of the problem (1.5.2). Let $x^{(0)} \in \mathcal{R}^n$ be an arbitrary initial approximation, and define the least-squares residual and error by*

$$
F(x) = \|Ax - b\|^2,
$$

$$
\delta(x) = F(x) - minF,
$$

then the RGSO method is linearly convergent in expectation to a solution in problem (1.5.2).

For each iteration:$k = 2, 3, \dots$,

$$
E_k\delta(x^{(k+1)}) \leqslant \left(1 - \frac{1}{(n-2)(k_F^2(A) - 1)}\right) \delta(x^{(k)}).
$$

In particular, if A has full column rank, we have the equivalent property

$$
E_k\left[\|x^{(k+1)} - x^*\|_{A^T A}^2\right] \leqslant \left(1 - \frac{1}{(n-2)(k_F^2(A) - 1)}\right) \|x^{(k)} - x^*\|_{A^T A}^2,
$$

where $x^* = A^\dagger b = (A^T A)^{-1} A^T b$ is the unique least-squares solution.

Proof. It is easy to prove that

$$
F(x) - F(\tilde{x}) = \|x - \tilde{x}\|_{A^T A}^2 = \delta(x).
$$

Applying the system (1.5.3) to the equation (1.2.2) with $A_{i_k}^T r_k = 0, k = 1, 2, ..$ and $d = e_{i_{k+1}} - \frac{\langle A_{i_{k+1}}, A_{i_k} \rangle}{||A_{i_k}||^2} e_{i_k}$, we get that

$$F(x^{(k+1)}) - F(\tilde{x}) = ||x^{(k+1)} - \tilde{x}||_{A^T A}^2$$

$$= ||x^{(k)} - \tilde{x}||_{A^T A}^2 - \frac{(A_{i_{k+1}}^T r^{(k)})^2}{g_{i_k}}.$$

Making conditional expectation on both sides, and applying Lemma 2.7.2, we get

$$E_k \left[F(x^{(k+1)}) - F(\tilde{x}) \right] = ||x^{(k)} - \tilde{x}||_{A^T A}^2 - E_k \left[\frac{(A_{i_{k+1}}^T r^{(k)})^2}{g_{i_k}} \right]$$

$$\leq ||x^{(k)} - \tilde{x}||_{A^T A}^2 - \frac{\sigma_{min}^2(A)||\tilde{x} - x^{(k)}||_{A^T A}^2}{(n-2)(||A||_F^2 - \sigma_{min}^2(A))},$$

that is

$$E_k \delta(x^{(k+1)}) \leq \left(1 - \frac{\sigma_{min}^2(A)}{(n-2)(||A||_F^2 - \sigma_{min}^2(A))} \right) \delta(x^{(k)})$$

$$= \left(1 - \frac{1}{(n-2)(k_F^2(A) - 1)} \right) \delta(x^{(k)}).$$

If A has full column rank, the solution in (1.1) is unique and the $\tilde{x} = x^*$. Thus, we get

$$E_k \left[||x^{(k+1)} - x^*||_{A^T A}^2 \right] \leq \left(1 - \frac{1}{(n-2)(k_F^2(A) - 1)} \right) ||x^{(k)} - x^*||_{A^T A}^2.$$

$\square$

Remark 2.7.2. *In particular, after unitizing the columns of matrix A, we can get from Lemma 2.7.2:*

$$E_k \frac{(A_{i_{k+1}}^T r^{(k)})^2}{g_{i_k}} = \frac{1}{n-2} \sum_{\substack{s=1 \\ s \neq i_k, i_{k-1}}}^{n} \frac{(A_s^T r^{(k)})^2}{||A_s||^2 - \frac{\langle A_s, A_{i_k} \rangle^2}{||A_{i_k}||^2}}$$

$$= \sum_{\substack{s=1 \\ s \neq i_k, i_{k-1}}}^{n} \frac{||A_s||^2}{||A||_F^2 - 2} \frac{(A_s^T r^{(k)})^2}{||A_s||^2 - \frac{\langle A_s, A_{i_k} \rangle^2}{||A_{i_k}||^2}}$$

$$\geqslant \sum_{\substack{s=1 \\ s \neq i_k, i_{k-1}}}^{n} \frac{1}{\|A\|_F^2 - 2} \frac{(A_s^T r^{(k)})^2}{1 - \gamma_{i_k}^2}$$

$$\geqslant \frac{\sigma_{min}^2(A)\|\tilde{x} - x^{(k)}\|_{A^T A}^2}{(1 - \gamma_{i_k}^2)(\|A\|_F^2 - 2)}, k = 2, 3, \dots$$

where $\gamma_{i_k} = \min\limits_{s \neq i_k, i_{k-1}} |\langle A_s, A_{i_k} \rangle|$. Then we get from Theorem 2.7.2:

$$E_k \delta(x^{(k+1)}) \leqslant \left(1 - \frac{\sigma_{min}^2(A)}{(1 - \gamma_{i_k}^2)(\|A\|_F^2 - 2)}\right) \delta(x^{(k)}).$$

Comparing the above equation with (2.5.3), we can get that under the condition of column unitization, the RGSO method is theoretically faster than the RCD method. When $g_{i_k} = 0$, $A_{i_{k+1}}$ is parallel to A_{i_k} (i.e. $\exists \lambda > 0$, s.t. $A_{i_k} = \lambda A_{i_{k+1}}$). According to the above derivation, the GSO method is used to solve (1.5.3) which is consistent, so the following equation holds:

$$A_{i_k}^T b = \lambda A_{i_{k+1}}^T b,$$

which means for (1.5.3) the i_kth equation: $\langle A_{i_k}, Ax \rangle = A_{i_k}^T b$, and the i_{k+1}th equation: $\langle A_{i_{k+1}}, Ax \rangle = A_{i_{k+1}}^T b$ are coincident, and we can skip this step without affecting the final calculation to obtain the least-squares solution. When g_{i_k} is too small, it is easy to produce large errors in the process of numerical operation, and we can regard it as the same situation as $g_{i_k} = 0$ and skip this step. So we can avoid the occurrence of $\gamma_{i_k} = 1$.

2.7.3   GRK and GRGS with Oblique Projection

Combining the oblique projection with the GRK method[36], we can obtain the GRKO method[51]. Theoretical results show that the KO-type method can accelerate the convergence when there are suitable row index selection strategies.

The core of the GRK method[36] is a new probability criterion, which can grasp the large items of the residual vector in each iteration, and randomly select the item with probability in proportion to the retained residual norm. Theories and experiments show these mathods can speed up the convergence speed. The GRKO algorithm is described in Algorithm 2.7.3

The convergence of the GRKO method is provided as follows.

Theorem 2.7.3. *Consider the consistent linear system (1.1.1), where the coefficient matrix $A \in \mathcal{R}^{m \times n}$, $b \in \mathcal{R}^m$. Let $x^{(0)} \in \mathcal{R}^n$ be an arbitrary initial approximation , $\tilde{x}$ is a solution of system (1.1.1) such that $P_{\mathcal{N}(A)}(\tilde{x}) = P_{\mathcal{N}(A)}(x^{(0)})$. Then the iteration sequence $\{x^{(k)}\}_{k=1}^{\infty}$ generated by the GRKO method obeys*

$$E||x^{(k)} - \tilde{x}||^2 \leqslant \prod_{s=0}^{k-1} \zeta_s ||x^{(0)} - \tilde{x}||^2. \tag{2.7.5}$$

where $\zeta_0 = 1 - \frac{(\lambda_{min}(A^T A))}{m||A||_F^2}$, $\zeta_1 = 1 - \frac{1}{2}(\frac{1}{\gamma_1}||A||_F^2 + 1)\frac{\lambda_{min}(A^T A)}{\Delta \cdot ||A||_F^2}$, $\zeta_k = 1 - \frac{1}{2}(\frac{1}{\gamma_2}||A||_F^2 + 1)\frac{\lambda_{min}(A^T A)}{\Delta \cdot ||A||_F^2}$ $(\forall k > 1)$, which

$$\gamma_1 = \max_{1 \leqslant i \leqslant m} \sum_{\substack{s=1 \\ s \neq i}}^{m} ||a_s||^2, \tag{2.7.6}$$

$$\gamma_2 = \max_{\substack{1 \leqslant i,j \leqslant m \\ i \neq j}} \sum_{\substack{s=1 \\ s \neq i,j}}^{m} ||a_s||^2, \tag{2.7.7}$$

$$\Delta = \max_{j \neq k} sin^2 \langle a_j, a_k \rangle (\in (0, 1]). \tag{2.7.8}$$

In addition, if $x^{(0)} \in \mathcal{R}(A^T)$, the sequence $\{x^{(k)}\}_{k=1}^{\infty}$ converges to the least-norm solution of the system (1.1.1), i.e. $\lim_{k \to \infty} x^{(k)} = x^* = A^{\dagger}b$.

Algorithm 2.7.3 GRK with Oblique Projection (GRKO)

procedure $(A,\ b,\ x^{(0)},\ K,\ M(i) = \|a_i\|^2,\ i \in [m])$

Uniformly randomly select i_1, and compute $x^{(1)} = x^{(0)} + \dfrac{b_{i_1} - \langle a_{i_1}, x^{(0)} \rangle}{M(i_1)} a_{i_1}$

for $k = 1, 2, ..., K - 1$ **do**

Compute $\varepsilon_k = \dfrac{1}{2} \left(\dfrac{1}{\|b - Ax^{(k)}\|^2} \max_{1 \leqslant i_{k+1} \leqslant m} \left\{ \dfrac{|b_{i_{k+1}} - \langle a_{i_{k+1}}, x^{(k)} \rangle|^2}{\|a_{i_{k+1}}\|^2} \right\} + \dfrac{1}{\|A\|_F^2} \right)$

Determine the index set of positive integers

$$\mathcal{U}_k = \left\{ i_{k+1} \big| |b_{i_{k+1}} - \langle a_{i_{k+1}}, x^{(k)} \rangle|^2 \geqslant \varepsilon_k \|b - Ax^{(k)}\|^2 \|a_{i_{k+1}}\|^2 \right\}$$

Compute the ith entry $\tilde{r}_i^{(k)}$ of the vector $\tilde{r}^{(k)}$ according to

$$\tilde{r}_i^{(k)} = \begin{cases} b_i - \langle a_i, x^{(k)} \rangle, & \text{if } i \in \mathcal{U}_k \\ 0 & \text{otherwise} \end{cases}$$

Select $i_{k+1} \in \mathcal{U}_k$ with probability $Pr(row = i_{k+1}) = \dfrac{|\tilde{r}_{i_{k+1}}^{(k)}|^2}{\|\tilde{r}^{(k)}\|^2}$

Compute $D_{i_k} = \langle a_{i_k}, a_{i_{k+1}} \rangle$

Compute $w_{i_k} = a_{i_{k+1}} - \dfrac{D_{i_k}}{M(i_k)} a_{i_k}$ and $h_{i_k} (= \|w_{i_k}\|^2) = M(i_{k+1}) - \dfrac{D_{i_k}^2}{M(i_k)}$

$\alpha_{i_k}^{(k)} = \dfrac{\tilde{r}_{i_{k+1}}^{(k)}}{h_{i_k}} \left(= \dfrac{r_{i_{k+1}}^{(k)}}{h_{i_k}} \right)$ and $x^{(k+1)} = x^{(k)} + \alpha_{i_k}^{(k)} w_{i_k}$

end for

Output $x^{(K)}$

end procedure

Proof. When $k = 1$, we can get

$$
\varepsilon_1 \|A\|_F^2 = \frac{\max\limits_{1 \leqslant i_2 \leqslant m} \left\{ \frac{|b_{i_2} - \langle a_{i_2}, x^{(1)} \rangle|^2}{\|a_{i_2}\|^2} \right\}}{2 \sum\limits_{i_2=1}^{m} \frac{\|a_{i_2}\|^2}{\|A\|_F^2} \cdot \frac{|b_{i_2} - \langle a_{i_2}, x^{(1)} \rangle|^2}{\|a_{i_2}\|^2}} + \frac{1}{2}
$$

$$
\overset{(iv)}{=} \frac{\max\limits_{1 \leqslant i_2 \leqslant m} \left\{ \frac{|b_{i_2} - \langle a_{i_2}, x^{(1)} \rangle|^2}{\|a_{i_2}\|^2} \right\}}{2 \sum\limits_{\substack{i_2=1 \\ i_2 \neq i_1}}^{m} \frac{\|a_{i_2}\|^2}{\|A\|_F^2} \cdot \frac{|b_{i_2} - \langle a_{i_2}, x^{(1)} \rangle|^2}{\|a_{i_2}\|^2}} + \frac{1}{2} \tag{2.7.9}
$$

$$
\geqslant \frac{1}{2} \left(\frac{\|A\|_F^2}{\sum\limits_{\substack{i_2=1 \\ i_2 \neq i_1}}^{m} \|a_{i_2}\|^2} + 1 \right)
$$

$$
\geqslant \frac{1}{2} \left(\frac{1}{\gamma_1} \|A\|_F^2 + 1 \right).
$$

The equality (iv) holds as $r_{i_k} = 0$.

When $k > 1$, we get

$$
\varepsilon_k \|A\|_F^2 = \frac{\max\limits_{1 \leqslant i_{k+1} \leqslant m} \left(\frac{|b_{i_{k+1}} - \langle a_{i_{k+1}}, x^{(k)} \rangle|^2}{\|a_{i_{k+1}}\|^2} \right)}{2 \sum\limits_{i_{k+1}=1}^{m} \frac{\|a_{i_{k+1}}\|^2}{\|A\|_F^2} \cdot \frac{|b_{i_{k+1}} - \langle a_{i_{k+1}}, x^{(k)} \rangle|^2}{\|a_{i_{k+1}}\|^2}} + \frac{1}{2}
$$

$$
\overset{(v)}{=} \frac{\max\limits_{1 \leqslant i_{k+1} \leqslant m} \left(\frac{|b_{i_{k+1}} - \langle a_{i_{k+1}}, x^{(k)} \rangle|^2}{\|a_{i_{k+1}}\|^2} \right)}{2 \sum\limits_{\substack{i_{k+1}=1 \\ i_{k+1} \neq i_k, i_{k-1}}}^{m} \frac{\|a_{i_{k+1}}\|^2}{\|A\|_F^2} \cdot \frac{|b_{i_{k+1}} - \langle a_{i_{k+1}}, x^{(k)} \rangle|^2}{\|a_{i_{k+1}}\|^2}} + \frac{1}{2} \tag{2.7.10}
$$

$$
\geqslant \frac{1}{2} \left(\frac{\|A\|_F^2}{\sum\limits_{\substack{i_{k+1}=1 \\ i_{k+1} \neq i_k, i_{k-1}}}^{m} \|a_{i_{k+1}}\|^2} + 1 \right)
$$

$$
\geqslant \frac{1}{2} \left(\frac{1}{\gamma_2} \|A\|_F^2 + 1 \right).
$$

The equality (v) holds as $r_{i_k} = 0$ and $r_{i_{k-1}} = 0$.

With the use of the orthogonality of $x^{(k+1)} - x^{(k)}$ and $x^{(k+1)} - \tilde{x}$, we can take the full expectation on both sides of the equation $\|x^{(k+1)} - \tilde{x}\|^2 =$

$||x^{(k)} - \tilde{x}||^2 - ||x^{(k+1)} - x^{(k)}||^2$ $(\forall k \geqslant 0)$, and get that for $k = 0$,

$$
\begin{aligned}
E||x^{(1)} - \tilde{x}||^2 &= ||x^{(0)} - \tilde{x}||^2 - E||x^{(1)} - x^{(0)}||^2 \\
&= ||x^{(0)} - \tilde{x}||^2 - \frac{1}{m} \sum_{i_1=1}^{m} ||\frac{b_{i_1} - \langle a_{i_1}, x^{(0)} \rangle}{M(i_1)} a_{i_1}||^2 \\
&\overset{(vi)}{\leqslant} ||x^{(0)} - \tilde{x}||^2 - \frac{1}{m} \frac{||b - Ax^{(0)}||^2}{||A||_F^2} \\
&\overset{(vii)}{\leqslant} \left(1 - \frac{\lambda_{min}(A^T A)}{m||A||_F^2}\right) ||x^{(0)} - \tilde{x}||^2 \\
&= \zeta_0 ||x^{(0)} - \tilde{x}||^2,
\end{aligned}
\tag{2.7.11}
$$

and for $k > 0$,

$$
\begin{aligned}
E_k||x^{(k+1)} - \tilde{x}||^2 &= ||x^{(k)} - \tilde{x}||^2 - E_k||x^{(k+1)} - x^{(k)}||^2 \\
&= ||x^{(k)} - \tilde{x}||^2 - \sum_{i_{k+1} \in \mathcal{U}_k} \frac{|b_{i_{k+1}} - \langle a_{i_{k+1}}, x^{(k)} \rangle|^2}{\sum_{i_{k+1} \in \mathcal{U}_k} |b_{i_{k+1}} - \langle a_{i_{k+1}}, x^{(k)} \rangle|^2} \cdot \frac{|r_{i_{k+1}}^{(k)}|^2}{||w_{i_k}||^2} \\
&\overset{(viii)}{\leqslant} ||x^{(k)} - \tilde{x}||^2 - \sum_{i_{k+1} \in \mathcal{U}_k} \frac{|b_{i_{k+1}} - \langle a_{i_{k+1}}, x^{(k)} \rangle|^2}{\sum_{i_{k+1} \in \mathcal{U}_k} |b_{i_{k+1}} - \langle a_{i_{k+1}}, x^{(k)} \rangle|^2} \cdot \frac{|r_{i_{k+1}}^{(k)}|^2}{\Delta \cdot ||a_{i_{k+1}}||^2} \\
&\overset{(ix)}{\leqslant} ||x^{(k)} - \tilde{x}||^2 - \frac{\varepsilon_k}{\Delta}||b - Ax^{(k)}||^2 \\
&= ||x^{(k)} - \tilde{x}||^2 - \frac{\varepsilon_k}{\Delta}||A(\tilde{x} - x^{(k)})||^2 \\
&\overset{(vii)}{\leqslant} \left(1 - \frac{\varepsilon_k \lambda_{min}(A^T A)}{\Delta}\right)||x^{(k)} - \tilde{x}||^2.
\end{aligned}
\tag{2.7.12}
$$

The inequality (vi) of the equation (2.7.11) is achieved with the use of the fact that $\frac{|b_1|}{|a_1|} + \frac{|b_2|}{|a_2|} \geqslant \frac{|b_1| + |b_2|}{|a_1| + |a_2|}$ (if $|a_1| > 0$, $|a_2| > 0$), and the inequality $(viii)$ of the equation (2.7.12) is achieved with the use of the fact that

$$
\begin{aligned}
||w_{i_k}||^2 &= ||a_{i_{k+1}}||^2 - \frac{\langle a_{i_k}, a_{i_{k+1}} \rangle^2}{||a_{i_{k+1}}||^2} \\
&= sin^2 \langle a_{i_k}, a_{i_{k+1}} \rangle ||a_{i_{k+1}}||^2 \\
&\leqslant \Delta \cdot ||a_{i_{k+1}}||^2,
\end{aligned}
\tag{2.7.13}
$$

and the inequality (ix) of the equation (2.7.12) is achieved with the use of the definition of $\mathcal{U}_k$ which leads to

$$
|b_{i_{k+1}} - \langle a_{i_{k+1}}, x^{(k)} \rangle|^2 \geqslant \varepsilon_k ||b - Ax^{(k)}||^2 ||a_{i_{k+1}}||^2, \forall i_{k+1} \in \mathcal{U}_k.
$$

Here in the last inequalities (vii) of the equation (2.7.11) and (2.7.12), we have used the estimate $||Au||_2^2 \geqslant \lambda_{min}(A^T A)||u||^2$, which holds true for any

$u \in R^n$ belonging to the column space of A^T. Using the orthogonality of $x^{(k+1)} - x^{(k)}$ and $x^{(k+1)} - \tilde{x}$, this relationship is established.

By making use of the equation (2.7.9), (2.7.10) and (2.7.12), we get

$$E_1||x^{(2)} - \tilde{x}||^2 \leqslant \left[1 - \frac{1}{2}(\frac{1}{\gamma_1}||A||_F^2 + 1)\frac{\lambda_{min}(A^T A)}{\Delta \cdot ||A||_F^2}\right]||x^{(1)} - \tilde{x}||^2$$
$$= \zeta_1||x^{(1)} - \tilde{x}||^2,$$

$$E_k||x^{(k+1)} - \tilde{x}||^2 \leqslant \left[1 - \frac{1}{2}(\frac{1}{\gamma_2}||A||_F^2 + 1)\frac{\lambda_{min}(A^T A)}{\Delta \cdot ||A||_F^2}\right]||x^{(k)} - \tilde{x}||^2$$
$$= \zeta_k||x^{(k)} - \tilde{x}||^2 \quad (\forall k > 1).$$

Finally, by recursion and taking the full expectation , the equation (2.7.5) holds. $\qquad\square$

Remark 2.7.3. *In the GRKO method, h_{i_k} is not zero. Suppose $h_{i_k} = 0$, which means $\exists \lambda > 0$, $\lambda a_{i_k} = a_{i_{k+1}}$. Since the system is consistent, it holds $\langle a_{i_{k+1}}, x^* \rangle = \lambda \langle a_{i_k}, x^* \rangle = \lambda b_{i_k} = b_{i_{k+1}}$. According to $r_{i_k} = 0$, it holds $r_{i_{k+1}}^{(k)} = \lambda r_{i_k}^{(k)} = 0$. From step 5 of Algorithm 2.7.1, we can know that such index i_{k+1} will not be selected.*

Remark 2.7.4. *Set $\tilde{\zeta}_k = 1 - \frac{1}{2}(\frac{1}{\gamma_1}||A||_F^2 + 1)\frac{\lambda_{min}(A^T A)}{||A||_F^2}$ $(\forall k > 0)$, and the convergence of GRK method in [36] meets:*

$$E_k||x^{(k+1)} - x^*||^2 \leqslant \tilde{\zeta}_k||x^{(k)} - x^*||^2.$$

Obviously, $\forall \Delta \in (0,1]$, $\zeta_1 \leqslant \tilde{\zeta}_1, \zeta_k < \tilde{\zeta}_k$ $(\forall k > 1)$ is satisfied, so the convergence speed of GRKO method is faster than GRK method. But, if $a_{i_{k+1}}$ is selected in the k-th iteration, and $\langle a_{i_k}, a_{i_{k+1}} \rangle = 0$, then $x^{(k+1)} = x^{(k)} + \alpha_{i_k}^{(k)} w_{i_k} = x^{(k)} + \frac{r_{i_{k+1}}^{(k)}}{||a_{i_{k+1}}||^2 - \frac{\langle a_{i_k}, a_{i_{k+1}} \rangle^2}{||a_{i_k}||^2}}(a_{i_{k+1}} - \frac{\langle a_{i_k}, a_{i_{k+1}} \rangle}{||a_{i_k}||^2} a_{i_k}) = x^{(k)} + \frac{r_{i_{k+1}}^{(k)}}{||a_{i_{k+1}}||^2} a_{i_{k+1}}$, which means the iteration degenerates to Kaczmarz-type iteration.

Remark 2.7.5. *In fact, in each iteration, the most computationally expensive part is computing the residual $r^{(k)}$. If $B = AA^T$ is calculated before iteration, the GRK method[36] costs $7m + 2n + 2$ flopping operations and the GRKO method costs $9m + 3n + 6$ flopping operations, where the residual $r^{(k)}$ is calculated according to recursive method.*

GRGS with Oblique Projection

Combining the oblique projection with the GRGS method[38], we can obtain the GRGSO method[52]. This method can be used to solve large-scale overdetermined inconsistent linear systems. When the coefficient matrix is of full column rank, the theoretical results show that the method converges to the unique least square solution of the linear system. Particularly, when the columns of matrix A are close to linearly correlated, the numerical results show that the method has more advantages than the RGS method. Due to space limitations, we have omitted the precise description and proof of this algorithm.

2.7.4   Maximal Weighted Residual KO and GSO Methods

The selection strategy for the index i_k used in the maximal weighted residual Kaczmarz (MWRK) method[53] is: the set

$$i_k = \arg\max_{i \in [m]} \frac{|a_i^T x^{(k)} - b_i|}{\|a_i\|}.$$

McCormick proved the exponential convergence of the MWRK method. In[54], a new convergence conclusion of the MWRK method was given. We use its row index selection rule combined with KO-type method to obtain MWRKO method[51], and the algorithm is described as follows:

The convergence of the MWRKO method is provided as follows.

Theorem 2.7.4. *Consider the consistent linear system (1.1.1), where the coefficient matrix $A \in \mathcal{R}^{m \times n}$, $b \in \mathcal{R}^m$. Let $x^{(0)} \in \mathcal{R}^n$ be an arbitrary initial approximation and $\tilde{x}$ be a solution of system (1.1.1) such that $P_{\mathcal{N}(A)}(\tilde{x}) =$*

Algorithm 2.7.4 Maximal Weighted Residual KO (MWRKO)

procedure $(A,\ b,\ x^{(0)},\ K,\ M(i) = \|a_i\|^2,\ i \in [m])$

Compute $i_1 = arg\max\limits_{i \in [m]} \dfrac{|a_i^T x^{(0)} - b_i|}{\|a_i\|}$, and $x^{(1)} = x^{(0)} + \dfrac{b_{i_1} - \langle a_{i_1}, x^{(0)} \rangle}{M(i_1)} a_{i_1}$

for $k = 1, 2, \cdots, K$ **do**

Compute $i_{k+1} = arg\max\limits_{i \in [m]} \dfrac{|a_i^T x^{(k)} - b_i|}{\|a_i\|}$

Compute $D_{i_k} = \langle a_{i_k}, a_{i_{k+1}} \rangle$ and $r_{i_{k+1}}^{(k)} = b_{i_{k+1}} - \langle a_{i_{k+1}}, x^{(k)} \rangle$

Compute $w_{i_k} = a_{i_{k+1}} - \dfrac{D_{i_k}}{M(i_k)} a_{i_k}$ and $h_{i_k} (= \|w_{i_k}\|^2) = M(i_{k+1}) - \dfrac{D_{i_k}^2}{M(i_k)}$

$\alpha_{i_k}^{(k)} = \dfrac{r_{i_{k+1}}^{(k)}}{h_{i_k}}$ and $x^{(k+1)} = x^{(k)} + \alpha_{i_k}^{(k)} w_{i_k}$

end for

Output $x^{(K+1)}$

end procedure

$P_{\mathcal{N}(A)}(x^{(0)})$. *Then the iteration sequence* $\{x^{(k)}\}_{k=1}^{\infty}$ *generated by the MWRKO method obeys*

$$\|x^{(k)} - \tilde{x}\|^2 \leqslant \prod_{s=0}^{k-1} \rho_s \|x^{(0)} - \tilde{x}\|^2, \qquad (2.7.14)$$

where $\rho_0 = 1 - \dfrac{\lambda_{min}(A^T A)}{\|A\|_F^2}$, $\rho_1 = 1 - \dfrac{\lambda_{min}(A^T A)}{\Delta \cdot \gamma_1}$, $\rho_k = 1 - \dfrac{\lambda_{min}(A^T A)}{\Delta \cdot \gamma_2}$ $(\forall k > 1)$, *which* γ_1, γ_2 *and* Δ *are defined by equations (2.7.6), (2.7.7) and (2.7.8) respectively.*

In addition, if $x^{(0)} \in \mathcal{R}(A^T)$, *the sequence* $\{x^{(k)}\}_{k=1}^{\infty}$ *converges to the least-norm solution of the system (1.1.1), i.e.* $\lim\limits_{k \to \infty} x^{(k)} = x^* = A^{\dagger}b$.

Proof. By the orthogonality of $x^{(1)} - x^{(0)}$ and $x^{(1)} - \tilde{x}$, we have

$$
\begin{aligned}
\|x^{(1)} - \tilde{x}\|^2 &= \|x^{(0)} - \tilde{x}\|^2 - \|x^{(1)} - x^{(0)}\|^2 \\
&= \|x^{(0)} - \tilde{x}\|^2 - \frac{|b_{i_1} - \langle a_{i_1}, x^{(0)}\rangle|^2}{M(i_1)} \\
&= \|x^{(0)} - \tilde{x}\|^2 - \frac{|b_{i_1} - \langle a_{i_1}, x^{(0)}\rangle|^2}{M(i_1)} \cdot \frac{\|b - Ax^{(0)}\|^2}{\sum\limits_{i=1}^{m} \frac{|b_i - \langle a_i, x^{(0)}\rangle|^2}{M(i)} \cdot M(i)} \\
&\overset{(x)}{\leqslant} \|x^{(0)} - \tilde{x}\|^2 - \frac{\|A(\tilde{x} - x^{(0)})\|^2}{\|A\|_F^2} \\
&\overset{(x)}{\leqslant} \|x^{(0)} - \tilde{x}\|^2 - \frac{\lambda_{min}(A^T A)}{\|A\|_F^2}\|x^{(0)} - \tilde{x}\|^2 \\
&= \left(1 - \frac{\lambda_{min}(A^T A)}{\|A\|_F^2}\right)\|x^{(0)} - x^*\|^2 \\
&= \rho_0 \|x^{(0)} - x^*\|^2.
\end{aligned}
\tag{2.7.15}
$$

Similarly, for $k = 1$, we have

$$
\begin{aligned}
\|x^{(2)} - \tilde{x}\|^2 &= \|x^{(1)} - \tilde{x}\|^2 - \|x^{(2)} - x^{(1)}\|^2 \\
&= \|x^{(1)} - \tilde{x}\|^2 - \frac{|b_{i_2} - \langle a_{i_2}, x^{(1)}\rangle|^2}{\|w_{i_1}\|^2} \\
&\overset{(xi)}{\leqslant} \|x^{(1)} - \tilde{x}\|^2 - \frac{|b_{i_2} - \langle a_{i_2}, x^{(1)}\rangle|^2}{\Delta \cdot M(i_2)} \cdot \frac{\|b - Ax^{(1)}\|^2}{\sum\limits_{i=1, i\neq i_1}^{m} \frac{|b_i - \langle a_i, x^{(1)}\rangle|^2}{M(i)} \cdot M(i)} \\
&\overset{(xii)}{\leqslant} \|x^{(1)} - \tilde{x}\|^2 - \frac{\|A(\tilde{x} - x^{(1)})\|^2}{\Delta \cdot \gamma_1} \\
&\overset{(x)}{\leqslant} \|x^{(1)} - \tilde{x}\|^2 - \frac{\lambda_{min}(A^T A)}{\Delta \cdot \gamma_1}\|x^{(1)} - \tilde{x}\|^2 \\
&= \left(1 - \frac{\lambda_{min}(A^T A)}{\Delta \cdot \gamma_1}\right)\|x^{(1)} - \tilde{x}\|^2 \\
&= \rho_1 \|x^{(1)} - \tilde{x}\|^2,
\end{aligned}
\tag{2.7.16}
$$

where the inequality (xi) can be obtained by using equations (2.7.13) and $r_{i_1}^{(1)} = 0$. For inequality (xii), using the row index selection rule of the MWRKO method, we get:

$$
\begin{aligned}
&\frac{|b_{i_2} - \langle a_{i_2}, x^{(1)}\rangle|^2}{\Delta \cdot M(i_2)} \cdot \frac{\|b - Ax^{(1)}\|^2}{\sum\limits_{i=1, i\neq i_1}^{m} \frac{|b_i - \langle a_i, x^{(1)}\rangle|^2}{M(i)} \cdot M(i)} \\
&= \max_{i \in \{1,2,\cdots,m\}} \frac{|b_i - \langle a_i, x^{(1)}\rangle|^2}{\Delta \cdot M(i)} \cdot \frac{\|b - Ax^{(1)}\|^2}{\sum\limits_{i=1, i\neq i_1}^{m} \frac{|b_i - \langle a_i, x^{(1)}\rangle|^2}{M(i)} \cdot M(i)} \\
&\geqslant \frac{\|b - Ax^{(1)}\|^2}{\Delta \cdot \sum\limits_{i=1, i\neq i_1}^{m} M(i)} \\
&\geqslant \frac{\|b - Ax^{(1)}\|^2}{\Delta \cdot \gamma_1}.
\end{aligned}
\tag{2.7.17}
$$

For $k > 1$, we have

$$\|x^{(k+1)} - \tilde{x}\|^2 = \|x^{(k)} - \tilde{x}\|^2 - \|x^{(k+1)} - x^{(k)}\|^2$$

$$= \|x^{(k)} - \tilde{x}\|^2 - \frac{|b_{i_{k+1}} - \langle a_{i_{k+1}}, x^{(k)}\rangle|^2}{\|w_{i_k}\|^2}$$

$$\overset{(xiii)}{\leqslant} \|x^{(k)} - \tilde{x}\|^2 - \frac{|b_{i_{k+1}} - \langle a_{i_{k+1}}, x^{(k)}\rangle|^2}{\Delta \cdot M(i_{k+1})} \cdot \frac{\|b - Ax^{(k)}\|^2}{\sum\limits_{i=1, i \neq i_k, i_{k-1}}^{m} \frac{|b_i - \langle a_i, x^{(k)}\rangle|^2}{M(i)} \cdot M(i)}$$

$$\overset{(xiv)}{\leqslant} \|x^{(k)} - \tilde{x}\|^2 - \frac{\|A(\tilde{x} - x^{(k)})\|^2}{\Delta \cdot \gamma_2}$$

$$\overset{(x)}{\leqslant} \|x^{(k)} - \tilde{x}\|^2 - \frac{\lambda_{min}(A^T A)}{\Delta \cdot \gamma_2} \|x^{(k)} - \tilde{x}\|^2$$

$$= \left(1 - \frac{\lambda_{min}(A^T A)}{\Delta \cdot \gamma_2}\right) \|x^{(k)} - \tilde{x}\|^2$$

$$= \rho_k \|x^{(k)} - \tilde{x}\|^2,$$

$$(2.7.18)$$

where the inequality $(xiii)$ can be obtained by using equations $(2.7.13)$, $r_{i_k}^{(k)} = 0$ and $r_{i_{k-1}}^{(k)} = 0$. For the inequality (xiv), it can be easily obtained by using a derivation similar to equation $(2.7.17)$. In the inequalities (x), we have used the estimate $\|Au\|_2^2 \geqslant \lambda_{min}(A^T A)\|u\|^2$, $\forall u \in \mathcal{R}(A^T)$.

From the equation $(2.7.15)$, $(2.7.16)$ and $(2.7.18)$, the equation $(2.7.14)$ holds. $\qquad\square$

Remark 2.7.6. *When multiple indicators i_{k+1} are met in Step 2 of Algorithm 2.7.3 in the iterative process, we randomly select any one of them.*

Remark 2.7.7. *In the MWRKO method, the reason of $h_{i_k} \neq 0$ is similar to Remark 2.7.3.*

Remark 2.7.8. *Set $\tilde{\rho}_0 = 1 - \frac{\lambda_{min}(A^T A)}{\|A\|_F^2}, \tilde{\rho}_k = 1 - \frac{\lambda_{min}(A^T A)}{\gamma_1}$ $(\forall k > 0)$, and the convergence of MWRK method in [54] meets:*

$$\|x^{(k)} - x^*\|^2 \leqslant \prod_{s=0}^{k-1} \tilde{\rho}_s \|x^{(0)} - x^*\|^2.$$

Obviously, $\forall \Delta \in (0, 1]$, $\rho_k < \tilde{\rho}_k$ $(\forall k > 1)$, $\rho_1 \leqslant \tilde{\rho}_1$ and $\rho_0 = \tilde{\rho}_0$, so the convergence speed of MWRKO method is faster than MWRK method. Note that $\tilde{\rho}_k < \tilde{\zeta}_k$, $\rho_k < \zeta_k$ $(\forall k > 0, \forall \Delta \in (0, 1])$, that is $V_{MWRK} < V_{MWRKO}$, $V_{GRK} < V_{GRKO}$, $V_{GRK} < V_{MWRK}$, $V_{GRKO} < V_{MWRKO}$, where V represents the convergence speed.

Maximal Weighted Residual GSO Method

Based on the greedy random Gauss-Seidel type methods with oblique direction and the working column selecting strategy of the max-distance for solving large-scale overdetermined inconsistent linear systems,[52] presented a FMDGSO method. When the coefficient matrix is of full column rank, the theoretical results show that the method converges to the unique least square solution of the linear system. Particularly, when the columns of matrix A are close to linearly correlated, the numerical results show that the method has more advantages than the RGS method. Due to space limitations, we have omitted the precise description and proof of this algorithm.

2.8   Residua-Based REK

As Bai and Wu said in [55], if the Euclidean norm of one column is much larger than the Euclidean norms of the others, Step 4 in the REK method would keep selecting this column at most of the time. In other words, if the Euclidean norm of one column is much smaller than the Euclidean norms of the others, Step 4 will not select this column at most of the time. In this case, some columns of the coefficient matrix A will fail to be chosen at most of the time.

When the coefficient matrix A is of full column rank, all its columns are independent of each other. If some of its columns are missed in the orthogonal projection process, the vector obtained in the second component of the REK method may not converge to $b_{\mathcal{R}(A)^\perp}$, so that the iteration sequence $\{x^{(k)}\}_{k=0}^\infty$ of the REK method may not converge to x^*. For example, for the linear system (1.1.1)with the coefficient matrix A and the right-hand side vector b being

$$A = \begin{pmatrix} 1 & 0 \\ 1 & \varepsilon \\ 1 & -\varepsilon \end{pmatrix}, \quad \begin{pmatrix} -2 \\ 2 \\ 0 \end{pmatrix}, \quad \varepsilon > 0. \tag{2.8.1}$$

we see that the matrix A is of full column rank, with the Euclidean norms of its two columns being $\sqrt{3}$ and $\sqrt{2}\varepsilon$, respectively, but this linear system is inconsistent as the rank of the corresponding augmented matrix $(A|b)$ is equal to 3. If ε is taken to be small enough, Step 4 in the REK method will keep selecting the first column of the matrix A. Then, since $A_1^T b = 0$, Step 5 will keep being $z^{(k+1)} = z^{(k)}$, so that the sequence $z^{(k+1)} = z^{(k)}$ will be not convergent to $b_{\mathcal{R}(A)^\perp}$. However, if we use the orthogonal projection in Step 5 in a given cyclic order (i.e., choose the column used in Step 5 as $A(1), A(2), A(1), A(2), ...$), then the sequence $\{z^{(k)}\}_{k=0}^{\infty}$ will converge to $b_{\mathcal{R}(A)^\perp}$ in only two steps of orthogonal projections.

While, if $\|a_i\|(i \in [m])$ is a constant, the PREK method[55] selects one row of matrix A with equal probability in the first component, then its advantage is lost. Motivated by[44,55], to solve the inconsistent system $Ax = b$ four extended Kaczmarz methods: two partially randomized methods like probability proportional to residual or residual homogenizing, and two deterministic strategies with the maximal-residual control and the maximum-distance control are proposed[56]. Without the full column rank and overdetermined assumptions on linear systems, a thorough convergence analysis in terms of expectation is derived.

2.8.1 Partially Randomized Extended Kaczmarz Method with Residuals

To solve inconsistent linear systems, a new partially random extended Kaczmarz method with residuals (PREKR) is established [56]. In this method, the probability of each row of the coefficient matrix A is proportional to the square of the Euclidean norm of the residual of each corresponding equation rather than the Euclidean norm of each row of matrix A in the first component, and the orthogonal projections in the second component are in the given cyclic order. Before the convergence is proved, the following lemmas firstly are given.

Algorithm 2.8.1 Partially REK with Residuals (PREKR)

    procedure $(A,\ b,\ l,\ x_0 \in \mathcal{R}(A^T)\backslash\{0\},\ z_0 = b)$

        for $k = 0, 1, \cdots, K - 1$ **do**

            Select $i_k \in [m]$ with probability $p_{i_k} = \dfrac{|r_{i_k}^{(k)}|^2}{\|r^{(k)}\|^2}$

            Set $x^{(k+1)} = x^{(k)} + \dfrac{b_{i_k} - z_{i_k}^{(k)} - \langle a_{i_k}, x^{(k)}\rangle}{\|a_{i_k}\|^2} a_{i_k}$

            Select $j_k = (k \bmod n) + 1$

            Set $z^{(k+1)} = z^{(k)} - \dfrac{\langle A_{j_k}, z^{(k)}\rangle}{\|A_{j_k}\|^2} A_{j_k}$

        end for

        Output $x^{(K)}$.

    end procedure

Lemma 2.8.1. *Let $x_1, \cdots, x_k$ be k $n-$dimensional vectors. Write $x_k = x_s + x_N$, where x_S belongs to the space S spanned by $x_1, \cdots, x_{k-1}$ and x_N is perpendicular to S. Let X be the matrix with columns $x_1, \cdots, x_{k-1}$. Then*

$$det(X, x_k)^T(X, x_k) = \|x_N\|^2 det(X^T X). \tag{2.8.2}$$

Proof. When $k > n$, $\hat{X} = (X, x_k)$ is underdetermined. Since x_N is perpendicular to S, it holds

$$\langle x_i, x_N \rangle = 0 (i = 1, 2, ...k - 1),$$

i.e.,

$$X^T x_N = 0. \tag{2.8.3}$$

There are two situations to consider:

    Case 1. $rank(X^T) = n$. From the equality (2.8.3), we have $x_N = 0$. Thus $x_k = x_s \in S$. With $S = span(x_1, x_2, ..., x_{k-1})$, we can assume that

$$x_k = c_1 x_1 + c_2 x_2 + ... + c_{k-1} x_{k-1},$$

where $c_i (i = 1, 2, ..., k - 1)$ are real numbers.

From

$$\hat{X}^T\hat{X} = \begin{pmatrix} x_1^T x_1 & x_1^T x_2 & \dots & x_1^T x_{k-1} & x_1^T x_k \\ x_2^T x_1 & x_2^T x_2 & \dots & x_2^T x_{k-1} & x_2^T x_k \\ . & . & \dots & . & . \\ x_k^T x_1 & x_k^T x_2 & \dots & x_k^T x_{k-1} & x_k^T x_k \end{pmatrix}, \quad (2.8.4)$$

we know that the last column of (2.8.4) can be represented linearly by the first k-1 columns. It follows that $det\hat{X}^T\hat{X} = 0$, then (2.8.2) is proved.

Case 2. $rank(X^T) < n$. From (2.8.3), it must exist $x_N \neq 0$ to make (2.8.3) true (If $x_N = 0$, it can be proved as in case 1.). Moreover, in this case $k - 1 \geqslant n$. Otherwise, $k - 1 < n$ and $n < k < n + 1$, such a k does not exist since k is an integer. Thus, in this case, $x_1, x_2, ..., x_{k-1}$ are linear correlation. Assume that x_1 can be represented linearly by $x_2, x_3, ..., x_{k-1}$, so that $det\left(\hat{X}^T\hat{X}\right) = 0$. Similarly, $det(X^TX) = 0$, (2.8.2) is followed.

When $k \leqslant n$, if $x_N = 0$, then $x_1, \cdots, x_k$ are dependent and both sides of (2.8.2) are 0. Otherwise let $e_0 = x_N/\|x_N\|$ and choose unit vectors $e_1, \cdots, e_{n-k}$ perpendicular to $x_1, \cdots, x_k$ and to each other. Then $x_k = \|x_N\|e_0 + c_1x_1 + \cdots + c_{k-1}x_{k-1}$, where $c_1, \cdots, c_{k-1}$ are scalars, which results in

$$\begin{aligned} det(X, x_k)^T(X, x_k) &= det^2(X, x_k, e_1, \cdots, e_{n-k}) \\ &= det^2(X, \|x_N\|e_0, e_1, \cdots, e_{n-k}) \\ &= \|x_N\|^2 det^2(X, e_0, e_1, \cdots, e_{n-k}) \\ &= \|x_N\|^2 det((X, e_0, e_1, \cdots, e_{n-k})^T(X, e_0, e_1, \cdots, e_{n-k})) \\ &= \|x_N\|^2 det \begin{pmatrix} X^TX & 0 \\ 0 & I \end{pmatrix} \\ &= \|x_N\|^2 det(X^TX). \end{aligned}$$

$\square$

Lemma 2.8.2. *Let $x_i \in \mathbb{R}^n$ ($i = 1, 2, \cdots, k$) be normalized vectors, where $k \geqslant 1$. S is the space spanned by these vectors. Assume that Q is the product*

$\left(I - x_k x_k^T\right) \ldots \left(I - x_2 x_2^T\right)\left(I - x_1 x_1^T\right)$, where I is the $n \times n$ identity matrix and x_i^T is the transpose of x_i. Denote by $X = (x_1, x_2, \ldots, x_k)$. Then

$$\max_{y \in S, \|y\|=1} \|Qy\| \leqslant (1 - det X^T X)^{1/2}. \tag{2.8.5}$$

Proof. When $k \leqslant n$, the proof has been given in Theorem 1 of[57]. Here it is generalized to $k \geqslant 1$. The method of proof is mathematical induction on k. The theorem is clearly true when $k = 1$, since both sides of (2.8.5) are 0. Suppose the theorem is true when $k = j - 1$, that is, suppose that

$$\max_{y \in S, \|y\|=1} \left\| \left(I - x_{j-1} x_{j-1}^T\right) \cdots \left(I - x_1 x_1^T\right) y \right\| \leqslant (1 - det X^T X)^{1/2}, \tag{2.8.6}$$

where now X is the matrix whose columns are $x_1, \cdots, x_{j-1}$ and S is the space spanned by $x_1, \cdots, x_{j-1}$. Write $x_j = x_N + x_S$, where x_S belongs to S and x_N is perpendicular to S. Let y be an arbitrary unit vector in the space spanned by $x_1, \cdots, x_{j-1}, x_j$, and write $y = y_N + y_S$, where y_S belongs to S and y_N is perpendicular to S. Finally, let $z = \left(I - x_{j-1} x_{j-1}^T\right) \cdots \left(I - x_1 x_1^T\right) y_S$. These vectors are represented in Fig.2.8.1. Using the facts that $\left(I - x_j x_j^T\right)^2 = \left(I - x_j x_j^T\right)$, that since x_N and y_N are parallel(or $y_N = 0$), $y_N^T x_N = \pm|y_N|\|x_N|$, and z is perpendicular to x_N and y_N (since it belongs to S), we then have

$$\left\| \left(I - x_{j-1} x_{j-1}^T\right) \cdots \left(I - x_1 x_1^T\right) y \right\|^2 \tag{2.8.7}$$
$$= \left\| \left(I - x_j x_j^T\right)(y_N + z) \right\|^2$$
$$= \left(y_N^T + z^T\right)\left(I - x_j x_j^T\right)(y_N + z)$$
$$= \|y_N\|^2 + \|z\|^2 - \left(y_N^T x_j\right)^2 - 2y_N^T x_j z^T x_j - \left(z^T x_j\right)^2$$
$$= \|y_N\|^2 + \|z\|^2 - \left(y_N^T x_j\right)^2 - 2y_N^T x_N z^T x_S - \left(z^T x_S\right)^2$$
$$= \|y_N\|^2 + \|z\|^2 - \|y_N\|^2\|x_N\|^2 \pm 2\|y_N\|\|x_N\| \left|z^T x_S\right| - \left\|z^T x_S\right\|^2$$
$$\leqslant \|y_N\|^2 + \|z\|^2 - \|y_N\|^2\|x_N\|^2 + 2\|y_N\|\|x_N\| \left|z^T x_S\right| - \left|z^T x_S\right|^2$$
$$= \|y_N\|^2 + \|z\|^2 - \left(\|y_N\|\|x_N\| - \left|z^T x_s\right|\right)^2.$$

There are now two cases to consider:

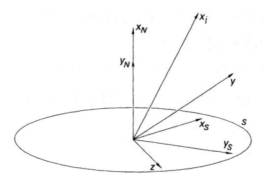

Figure 2.8.1: Lemma 2.8.1

Case 1. $\|x_N\| \leqslant \|y_S\|$. With the use of (2.8.7), the induction hypothesis (2.8.6), which says that $\|z\| \leqslant \|y_S\| \left(1 - detX^TX\right)^{1/2}$, and Lemma 2.8.1, then we have

$$
\begin{aligned}
\left\|\left(I - x_j x_j^T\right) \cdots \left(I - x_1 x_1^T\right) y\right\|^2 &\leqslant \|y_N\|^2 + \|y_S\|^2 \left(1 - detX^TX\right) \\
&\leqslant 1 - \|x_N\|^2 det(X^TX) \\
&= 1 - det(X, x_j)^T (X, x_j).
\end{aligned}
$$

Case 2. $\|x_N\| > \|y_S\|$. $\|y_N\| = \left(1 - \|y_s\|^2\right)^{1/2} > (1 - \|x_N\|^2)^{1/2} = \|x_S\|$, so that

$$
\|y_N\|\|x_N\| > \|y_S\|\|x_S\| \geqslant \|z\|\|x_S\| \geqslant |z^T x_S|
$$

and hence

$$
\|y_N\|\|x_N\| - |z^T x_S| \geqslant \|y_N\|\|x_N\| - \|z\|\|x_S\| \geqslant 0. \tag{2.8.8}
$$

With the use of (2.8.7), (2.8.8), the induction hypothesis (2.8.6), Cauchy's inequality, and Lemma 2.8.1, in that order, we then have

$$
\begin{aligned}
&\left\|\left(I - x_{j-1} x_{j-1}^T\right) \cdots \left(I - x_1 x_1^T\right) y\right\|^2 \\
&\leqslant \|y_N\|^2 + \|z\|^2 - (\|y_N\|\|x_N\| - \|z\|\|x_S\|)^2 \\
&= \|y_N\|(1 - \|x_N\|^2) + \|z\|^2(1 - \|x_S\|^2) + 2\|y_N\|\|x_N\|\|z\|\|x_S\|
\end{aligned}
$$

$$=\|y_N\|\|x_S\|^2 + \|z\|^2\|x_N\|^2 + 2\|y_N\|\|x_N\|\|z\|\|x_S\|$$

$$=(\|y_N\|\|x_S\| + \|z\|\|x_N\|)^2$$

$$\leqslant(\|y_N\|\|x_S\| + \|y_S\|\|x_N\|(1 - det(X^TX))^{1/2})^2$$

$$\leqslant(\|y_N\|^2 + \|y_S\|^2)(\|x_S\|^2 + \|x_N\|^2(1 - det(X^TX)))$$

$$=1 - det(X, x_j)^T(X, x_j).$$

$\square$

Lemma 2.8.3 ([58]). *Let α_1, β_1 be real numbers such that*

$$\alpha_1 \in [0,1), \quad \beta_1 \geqslant -1 \quad and \quad \beta_1 - \alpha_1 = \alpha_1\beta_1. \tag{2.8.9}$$

Then

$$(r_1 + r_2)^2 \geqslant \alpha_1 r_1^2 - \beta_1 r_2^2, \quad \forall r_1, r_2 \in \mathbb{R}. \tag{2.8.10}$$

This gives us the following result.

Lemma 2.8.4 ([58]). *Let α_1, β_1 be as in (2.8.9). Then*

$$\|x + y\|^2 \geqslant \alpha_1\|x\|^2 - \beta_1\|y\|^2, \quad \forall x, y \in \mathbb{R}^n. \tag{2.8.11}$$

Lemma 2.8.5. *[55] Let $A \in \mathbb{R}^{m \times n}$ be the coefficient matrix of the linear system. $\{z^{(k)}\}_{k=0}^{\infty}$ is a sequence generated by the PREKR method with $x^{(0)} \in \mathcal{R}(A^T)\backslash\{0\}$. Then, for any integer $k \geqslant 0$, the following relations are true:*

$$(1) \quad z^{(k)} - b_{\mathcal{R}(A)^\perp} \in \mathcal{R}(A) = \mathcal{R}(\tilde{A}). \tag{2.8.12}$$

$$(2) \quad \|z^{(k+n)} - b_{\mathcal{R}(A)^\perp}\|^2 \leqslant \beta\|z^{(k)} - b_{\mathcal{R}(A)^\perp}\|^2, \quad \beta = 1 - \frac{det(A^TA)}{\prod\limits_{j=1}^{n}\|A_j\|^2}. \tag{2.8.13}$$

$$(3) \quad \|z^{(k)} - b_{\mathcal{R}(A)^\perp}\|^2 \leqslant \|b_{\mathcal{R}(A)}\|^2. \tag{2.8.14}$$

Next, we can establish the convergence theory for the PREKR method.

Theorem 2.8.1. *Let $A \in \mathbb{R}^{m \times n}$ be the coefficient matrix of the linear system, α_1, β_1 be defined as in (2.8.9). Then, the iteration sequence $\{x^{(k)}\}_{k=0}^{\infty}$*

generated by the PREKR method with $x^{(0)} \in \mathcal{R}(A^T)\backslash\{0\}$ converges to the least-squares solution $x^* = A^\dagger b$ in expectation. Moreover, the solution error in expectation for the iteration sequence $\{x^{(k)}\}_{k=0}^{\infty}$ obeys

$$E\left\|x^{(k)} - x^*\right\|^2 \leqslant \alpha^k \|x^{(0)} - x^*\|^2 + \frac{\left(\alpha^{k-\lfloor\frac{k}{2n}\rfloor \cdot n} + \beta^{\lfloor\frac{k}{2n}\rfloor}\right)\mu\gamma\kappa^2(A)}{\alpha_1^2}\|x^*\|^2,$$

$$(2.8.15)$$

where

$$\alpha = 1 - \frac{\alpha_1^2\lambda_{min}(A^TA)}{\gamma}, \quad \beta = 1 - \frac{det(A^TA)}{\prod\limits_{j=1}^{n}\|A_j\|^2}, \quad \mu = \frac{\alpha_1\beta_1}{\gamma} + \frac{1+\beta_1}{\lambda_{min}(A^TA)}$$

with

$$\gamma = \|A\|_F^2 - \min_{1\leqslant i\leqslant m}\|a_i\|^2.$$

Proof. Firstly, we show the fact that in the PREKR method, it holds

$$r_{i_{k-1}}^{(k)} = b_{i_{k-1}} - \langle a_{i_{k-1}}, x^{(k)}\rangle - z_{i_{k-1}}^{(k)}$$

$$= b_{i_{k-1}} - \left\langle a_{i_{k-1}}, x^{(k-1)} + \frac{b_{i_{k-1}} - z_{i_{k-1}}^{(k-1)} - \langle a_{i_{k-1}}, x^{(k-1)}\rangle}{\|a_{i_{k-1}}\|^2}a_{i_{k-1}}\right\rangle - z_{i_{k-1}}^{(k)}$$

$$= b_{i_{k-1}} - \langle a_{i_{k-1}}, x^{(k-1)}\rangle - \left(b_{i_{k-1}} - z_{i_{k-1}}^{(k-1)} - \langle a_{i_{k-1}}, x^{(k-1)}\rangle\right) - z_{i_{k-1}}^{(k)}$$

$$= 0, \quad k = 1, 2, \ldots.$$

Now, substituting the direct-sum decomposition $b = b_{\mathcal{R}(A)^\perp} + b_{\mathcal{R}(A)}$ for the right-hand side b and the equality $b_{\mathcal{R}(A)} = Ax_*$ into the iteration scheme of Algorithm 2.8.1, we have

$$x^{(k+1)} - x^* = x^{(k)} - x^* + \frac{b_{\mathcal{R}(A)}^{(i_k)} - \langle a_{i_k}, x^{(k)}\rangle}{\|a_{i_k}\|^2}a_{i_k} + \frac{(b_{\mathcal{R}(A)^\perp})_{i_k} - z_{i_k}^{(k)}}{\|a_{i_k}\|^2}a_{i_k}$$

$$= \left(I_n - \frac{a_{i_k}(a_{i_k})^T}{\|a_{i_k}\|^2}\right)(x^{(k)} - x^*) + \frac{(b_{\mathcal{R}(A)^\perp})_{i_k} - z_{i_k}^{(k)}}{\|a_{i_k}\|^2}a_{i_k}.$$

Since the two terms on the right side of the equation are perpendicular to each other, it holds that

$$\|x^{(k+1)} - x^*\|^2 = \left\|\left(I_n - \frac{a_{i_k}(a_{i_k})^T}{\|a_{i_k}\|^2}\right)(x^{(k)} - x^*)\right\|^2 + \left\|\frac{(b_{\mathcal{R}(A)^\perp})_{i_k} - z_{i_k}^{(k)}}{\|a_{i_k}\|^2}a_{i_k}\right\|^2,$$

$$(2.8.16)$$

where the vector x_* is the least-norm least-squares solution of the system $Ax = b$. By taking the expectation conditional on the first k iterations on both sides of (2.8.16) and using the linearity of the expectation, we have

$$
\mathbb{E}_k \|x^{(k+1)} - x^*\|^2
$$

$$
= \mathbb{E}_k \left\| \left(I_n - \frac{a_{i_k}(a_{i_k})^T}{\|a_{i_k}\|^2} \right) (x^{(k)} - x^*) \right\|^2 + \mathbb{E}_k \left\| \frac{(b_{\mathcal{R}(A)^\perp})_{i_k} - z_{i_k}^{(k)}}{\|a_{i_k}\|^2} a_{i_k} \right\|^2
$$

$$
= \sum_{i_k=1}^m \frac{|r_{i_k}^{(k)}|^2}{\|r^{(k)}\|^2} (x^{(k)} - x^*)^T \left(I_n - \frac{a_{i_k}(a_{i_k})^T}{\|a_{i_k}\|^2} \right) (x^{(k)} - x^*)
$$

$$
+ \sum_{i_k=1}^m \frac{|r_{i_k}^{(k)}|^2}{\|r_k\|^2} \frac{\left|(b_{\mathcal{R}(A)^\perp})_{i_k} - z_{i_k}^{(k)}\right|^2}{\|a_{i_k}\|^2}
$$

$$
= \|x^{(k)} - x^*\|^2 - \sum_{i_k=1}^m \frac{|r_{i_k}^{(k)}|^2}{\|r^{(k)}\|^2} (x^{(k)} - x^*)^T \frac{a_{i_k}(a_{i_k})^T}{\|a_{i_k}\|^2} (x^{(k)} - x^*)
$$

$$
+ \sum_{i_k=1}^m \frac{|r_{i_k}^{(k)}|^2}{\|r_k\|^2} \frac{\left|(b_{\mathcal{R}(A)^\perp})_{i_k} - z_{i_k}^{(k)}\right|^2}{\|a_{i_k}\|^2}
$$

$$
= \|x^{(k)} - x^*\|^2 - \sum_{i_k=1}^m \frac{|r_{i_k}^{(k)}|^2}{\|r^{(k)}\|^2} \frac{|\langle a_{i_k}, x^{(k)} - x^*\rangle|^2}{\|a_{i_k}\|^2} + \sum_{i_k=1}^m \frac{|r_k^{(i_k)}|^2}{\|r^{(k)}\|^2} \frac{\left|(b_{\mathcal{R}(A)^\perp})_{(i_k)} - z_{i_k}^{(k)}\right|^2}{\|a_{i_k}\|^2}
$$

$$
= \|x^{(k)} - x^*\|^2 - \sum_{i_k=1}^m \frac{|r_{i_k}^{(k)}|^2}{\|r^{(k)}\|^2} \frac{|(b_{\mathcal{R}(A)^\perp})_{i_k} - \langle a_{i_k}, x^{(k)}\rangle|^2}{\|a_{i_k}\|^2}
$$

$$
+ \sum_{i_k=1}^m \frac{|r_{i_k}^{(k)}|^2}{\|r^{(k)}\|^2} \frac{\left|(b_{\mathcal{R}(A)^\perp})_{(i_k)} - z_k^{(i_k)}\right|^2}{\|a_{i_k}\|^2}
$$

$$
= \|x^{(k)} - x^*\|^2 + \sum_{i_k=1}^m \frac{|r_{i_k}^{(k)}|^2}{\|r^{(k)}\|^2} \frac{\left|(b_{\mathcal{R}(A)^\perp})_{i_k} - z_{i_k}^{(k)}\right|^2}{\|a_{i_k}\|^2}
$$

$$
- \sum_{i_k=1}^m \frac{|r_{i_k}^{(k)}|^2}{\|r^{(k)}\|^2} \frac{\left|(b_{\mathcal{R}(A)})_{i_k} + (b_{\mathcal{R}(A)^\perp})_{i_k} - \langle a_{i_k}, x^{(k)}\rangle - z_{i_k}^{(k)} + z_{i_k}^{(k)} - (b_{\mathcal{R}(A)^\perp})_{i_k}\right|^2}{\|a_{i_k}\|^2}
$$

$$
\leqslant \|x^{(k)} - x^*\|^2 - \sum_{i_k=1}^m \frac{|r_{i_k}^{(k)}|^2}{\|r^{(k)}\|^2} \frac{\alpha_1 \left|b^{(i_k)} - \langle a_{i_k}, x^{(k)}\rangle - z_{i_k}^{(k)}\right|^2}{\|a_{i_k}\|^2}
$$

$$
+ \sum_{i_k=1}^m \frac{|r_{i_k}^{(k)}|^2}{\|r^{(k)}\|^2} \frac{\beta_1 \left|(b_{\mathcal{R}(A)^\perp})_{i_k} - z_{i_k}^{(k)}\right|^2}{\|a_{i_k}\|^2} + \sum_{i_k=1}^m \frac{|r_k^{(i_k)}|^2}{\|r^{(k)}\|^2} \frac{\left|(b_{\mathcal{R}(A)^\perp})_{i_k} - z_{i_k}^{(k)}\right|^2}{\|a_{i_k}\|^2}
$$

$$
= \|x^{(k)} - x^*\|^2 - \alpha_1 \sum_{i_k=1}^m \frac{|r_{i_k}^{(k)}|^2}{\|r^{(k)}\|^2} \frac{|r_{i_k}^{(k)}|^2}{\|a_{i_k}\|^2}
$$

$$
+ (1 + \beta_1) \sum_{i_k=1}^m \frac{|r_{i_k}^{(k)}|^2}{\|r^{(k)}\|^2} \frac{\left|(b_{\mathcal{R}(A)^\perp})_{i_k} - z_{i_k}^{(k)}\right|^2}{\|a_{i_k}\|^2}.
$$

$$(2.8.17)$$

The inequality is obtained by (2.8.10). To estimate the bounds, we divide the right of the above equation into three parts.

For the second term, we have

$$\sum_{i_k=1}^{m} \frac{|r_k^{(i_k)}|^2}{\|r^{(k)}\|^2} \frac{|r_{i_k}^{(k)}|^2}{\|a_{i_k}\|^2} = \sum_{\substack{i_k=1 \\ i_k \neq i_{k-1}}}^{m} \frac{1}{\|r^{(k)}\|^2} \frac{\left(|r_{i_k}^{(k)}|^2\right)^2}{\|a_{i_k}\|^2}$$

$$\geq \frac{1}{\|r^{(k)}\|^2} \frac{\left(\sum\limits_{\substack{i_k=1 \\ i_k \neq i_{k-1}}}^{m} |r_{i_k}^{(k)}|^2\right)^2}{\sum\limits_{\substack{i_k=1 \\ i_k \neq i_{k-1}}}^{m} \|a_{i_k}\|^2}$$

$$\geq \frac{\|r^{(k)}\|^2}{\gamma}$$

$$= \frac{\|b - Ax^{(k)} - z^{(k)}\|^2}{\gamma}$$

$$\geq \frac{\alpha_1 \|b_{\mathcal{R}(A)} - Ax^{(k)}\|^2}{\gamma} - \frac{\beta_1 \|b_{\mathcal{R}(A)^\perp} - z^{(k)}\|^2}{\gamma}$$

$$= \frac{\alpha_1 \|A(x^{(k)} - x^*)\|^2}{\gamma} - \frac{\beta_1 \|b_{\mathcal{R}(A)^\perp} - z^{(k)}\|^2}{\gamma}$$

$$\geq \frac{\alpha_1 \lambda_{min}(A^T A)}{\gamma} \|x^{(k)} - x^*\|^2 - \frac{\beta_1 \|b_{\mathcal{R}(A)^\perp} - z^{(k)}\|^2}{\gamma}.$$

Here, the first equality is valid as $r_{i_{k-1}}^{(k)} = 0$, the first and third inequalities are achieved with the use of Lemmas 2.3.1 and 2.8.4, respectively. For the last inequality, the following estimate

$$\|Az\|^2 \geq \lambda_{min}(A^T A)\|z\|^2$$

is used. Since $x^{(k)} - x^* \in \mathcal{R}(A)$, the above inequality holds.

For the third part, since $b_{\mathcal{R}(A)^\perp} - z^{(k)} \in \mathbb{R}(A)$, there exists $\hat{x}$ which satisfies $b_{\mathcal{R}(A)^\perp} - z^{(k)} = A\hat{x}$. Let $\hat{x} = A^\dagger(b_{\mathcal{R}(A)^\perp} - z^{(k)})$, it holds

$$\sum_{i_k=1}^{m} \frac{|r_{i_k}^{(k)}|^2}{\|r^{(k)}\|^2} \frac{\left|(b_{\mathcal{R}(A)^\perp})_{i_k} - z_{i_k}^{(k)}\right|^2}{\|a_{i_k}\|^2}$$

$$= \sum_{\substack{i_k=1 \\ i_k \neq i_{k-1}}}^{m} \frac{|r_{i_k}^{(k)}|^2}{\|r^{(k)}\|^2} \frac{\left|(b_{\mathcal{R}(A)^\perp})_{i_k} - z_{i_k}^{(k)}\right|^2}{\|a_{i_k}\|^2}$$

$$= \sum_{\substack{i_k=1 \\ i_k \neq i_{k-1}}}^{m} \frac{|(b_{\mathcal{R}(A)})_{i_k} - \langle a_{i_k}, x^{(k)} \rangle + (b_{\mathcal{R}(A)^\perp})_{i_k} - z_{i_k}^{(k)}|^2}{\|r^{(k)}\|^2} \frac{\left|(b_{\mathcal{R}(A)^\perp})_{i_k} - z_{i_k}^{(k)}\right|^2}{\|a_{i_k}\|^2}$$

$$= \sum_{\substack{i_k=1 \\ i_k \neq i_{k-1}}}^{m} \frac{|\langle a_{i_k}, x^* - x^{(k)} - \hat{x} \rangle|^2}{\|r_k\|^2} \frac{\left|(b_{\mathcal{R}(A)^\perp})_{i_k} - z_{i_k}^{(k)}\right|^2}{\|a_{i_k}\|^2}$$

$$\leqslant \sum_{i_k=1}^{m} \frac{\|a_{i_k}\|^2 \|x^* - x^{(k)} - \hat{x}\|^2}{\|r^{(k)}\|^2} \frac{\left|(b_{\mathcal{R}(A)^\perp})_{i_k} - z_{i_k}^{(k)}\right|^2}{\|a_{i_k}\|^2}$$

$$= \sum_{i_k=1}^{m} \frac{\|x^* - x^{(k)} - \hat{x}\|^2}{\|A(x^* - x^{(k)} - \hat{x})\|^2} \left|(b_{\mathcal{R}(A)^\perp})_{i_k} - z_{i_k}^{(k)}\right|^2$$

$$= \sum_{i_k=1}^{m} \frac{1}{\frac{\|A(x^* - x^{(k)} - \hat{x})\|^2}{\|x^* - x^{(k)} - \hat{x}\|^2}} \left|(b_{\mathcal{R}(A)^\perp})_{i_k} - z_{i_k}^{(k)}\right|^2$$

$$\leqslant \frac{1}{\lambda_{min}(A^T A)} \sum_{i_k=1}^{m} \left|(b_{\mathcal{R}(A)^\perp})_{i_k} - z_{i_k}^{(k)}\right|^2$$

$$= \frac{1}{\lambda_{min}(A^T A)} \left\|b_{\mathcal{R}(A)^\perp} - z^{(k)}\right\|^2.$$

Therefore

$$\mathbb{E}_k \|x^{(k+1)} - x^*\|^2 \leqslant \alpha \|x^{(k)} - x^*\|^2 + \mu \left\|b_{\mathcal{R}(A)^\perp} - z^{(k)}\right\|^2, \qquad (2.8.18)$$

where $\alpha = 1 - \frac{\alpha_1^2 \lambda_{min}(A^T A)}{\gamma}$, $\mu = \frac{\alpha_1 \beta_1}{\gamma} + \frac{1+\beta_1}{\lambda_{min}(A^T A)}$.

For any nonnegative integer k, define $k_1 := \lfloor \frac{k}{2n} \rfloor \cdot n$ and $k_2 := k - k_1$, then from (2.8.18) we have

$$\mathbb{E} \|x^{(k_1-1)} - x^*\|^2$$

$$\leqslant \alpha \mathbb{E} \|x^{(k_1-1)} - x^*\|^2 + \mu \|b_{\mathcal{R}(A)^\perp} - z^{(k_1-1)}\|^2$$

$$\leqslant \alpha^2 \mathbb{E} \|x^{(k_1-2)} - x^*\|^2 + \alpha \mu \|b_{\mathcal{R}(A)^\perp} - z^{(k_1-2)}\|^2 + \mu \|b_{\mathcal{R}(A)^\perp} - z^{(k_1-1)}\|^2$$

$$\leqslant \cdots$$

$$\leqslant \ \alpha^{k_1}\|x^{(0)} - x^*\|^2 + \sum_{t=0}^{k_1-1} \alpha^l \mu \|b_{\mathcal{R}(A)^\perp} - z^{(k_1-l-1)}\|^2.$$

In addition, since $0 \leqslant \alpha < 1$, with (2.8.14) we can get

$$\mathbb{E}\|x^{(k_1)} - x^*\|^2 \leqslant \alpha^{k_1}\|x^{(0)} - x^*\|^2 + \mu\|b_{\mathcal{R}(A)}\|^2 \sum_{l=0}^{k_1-1} \alpha^l$$

$$\leqslant \alpha^{k_1}\|x^{(0)} - x^*\|^2 + \mu\|b_{\mathcal{R}(A)}\|^2 \sum_{l=0}^{\infty} \alpha^l \qquad (2.8.19)$$

$$= \alpha^{k_1}\|x^{(0)} - x^*\|^2 + \frac{\mu\|b_{\mathcal{R}(A)}\|^2}{1 - \alpha}.$$

For any nonnegative integer k, it follows from (2.8.18) again that

$$\mathbb{E}\|x^{(k)} - x^*\|^2 = \mathbb{E}\|x^{(k_1+k_2)} - x^*\|^2$$

$$\leqslant \alpha\mathbb{E}\|x^{(k_1+k_2-1)} - x^*\|^2 + \mu\|b_{\mathcal{R}(A)^\perp} - z^{(k_1+k_2-1)}\|^2$$

$$\leqslant \cdots$$

$$\leqslant \alpha^{k_2}\mathbb{E}\|x^{(k_1)} - x^*\|^2 + \sum_{l=0}^{k_2-1} \alpha^l \mu \|b_{\mathcal{R}(A)^\perp} - z^{(k_1+k_2-l-1)}\|^2.$$

With the substitutions of (2.8.13), (2.8.14) and $k_1 = \lfloor \frac{k}{2n} \rfloor \cdot n$, we can obtain the estimate

$$\mathbb{E}\|x^{(k)} - x^*\|^2 \leqslant \alpha^{k_2}\mathbb{E}\|x^{(k_1)} - x^*\|^2 + \beta^{\lfloor \frac{k}{2n} \rfloor} \sum_{l=0}^{k_2-1} \alpha^l \mu \|b_{\mathcal{R}(A)^\perp} - z^{(k_2-l-1)}\|^2$$

$$\leqslant \alpha^{k_2}\mathbb{E}\|x^{(k_1)} - x^*\|^2 + \beta^{\lfloor \frac{k}{2n} \rfloor}\mu\|b_{\mathcal{R}(A)}\|^2 \sum_{l=0}^{k_2-1} \alpha^l$$

$$\leqslant \alpha^{k_2}\mathbb{E}\|x^{(k_1)} - x^*\|^2 + \beta^{\lfloor \frac{k}{2n} \rfloor}\mu\|b_{\mathcal{R}(A)}\|^2 \sum_{l=0}^{\infty} \alpha^l$$

$$= \alpha^{k_2}\mathbb{E}\|x^{(k_1)} - x^*\|^2 + \beta^{\lfloor \frac{k}{2n} \rfloor}\frac{\mu\|b_{\mathcal{R}(A)}\|^2}{1 - \alpha}.$$

Here we have used the fact $0 \leqslant \alpha < 1$ again. Making use of the inequality (2.8.19) and the bound

$$\|b_{\mathcal{R}(A)}\|^2 \leqslant \lambda_{max}(A^T A)\|x^*\|^2,$$

we know that

$$\mathbb{E}\|x^{(k)} - x^*\|^2 \leq \alpha^{k_2}\left(\alpha^{k_1}\|x^{(0)} - x^*\|^2 + \frac{\mu\|b_{\mathcal{R}(A)}\|^2}{1-\alpha}\right) + \beta^{\lfloor\frac{k}{2n}\rfloor}\frac{\mu\|b_{\mathcal{R}(A)}\|^2}{1-\alpha}$$

$$\leq \alpha^k\|x^{(0)} - x^*\|^2 + \left(\alpha^{k_2} + \beta^{\lfloor\frac{k}{2n}\rfloor}\right)\mu\gamma\kappa^2(A)\|x^*\|^2/\alpha_1^2,$$

which is exactly the estimate (2.8.15) with

$$k_2 = k - k_1 = k - \lfloor\frac{k}{2n}\rfloor \cdot n.$$

$\square$

Remark 2.8.1. It can be seen from the proof of Theorem 2.8.1 that if all the columns of the coefficient matrix A are orthogonal to each other, the iterative subsequence $\{z^{(k)}\}_{k=0}^{\infty}$ in the PREKR method will converge in n steps. From Algorithm 2.8.1, we can get

$$z^{(n)} - b_{\mathcal{R}(A)^{\perp}}$$
$$= \left(I_m - \frac{A_n A_n^T}{\|A_n\|^2}\right)\left(I_m - \frac{A_{n-1}A_{n-1}^T}{\|A_{n-1}\|^2}\right)\cdots\left(I_m - \frac{A_1 A_1^T}{\|A_1\|^2}\right)\cdot\left(z^{(0)} - b_{\mathcal{R}(A)^{\perp}}\right)$$
$$= \left(I_m - \frac{A_n A_n^T}{\|A_n\|^2} - \frac{A_{n-1}A_{n-1}^T}{\|A_{n-1}\|^2} - \cdots - \frac{A_1 A_1^T}{\|A_1\|^2}\right)b_{\mathcal{R}(A)} = 0.$$

Meanwhile, Theorem 2.8.1 shows that in this case, it results in

$$r^{(k)} = b - Ax^{(k)} - z^{(k)}$$
$$= b_{\mathcal{R}(A)} - Ax^{(k)} - (z^{(n)} - b_{\mathcal{R}(A)^{\perp}})$$
$$= b_{\mathcal{R}(A)} - Ax^{(k)}, \quad k \geq n.$$

Since $r_{i_k}^{(k)} = \langle a_{i_k}, x^* - x^{(k)}\rangle (k \geq n)$ in (2.8.17), it follows

$$\tilde{\alpha} = 1 - \frac{\lambda_{min}(A^T A)}{\|A\|_F^2 - \min\limits_{1\leq i\leq m}\|a_i\|^2} \leq 1 - \frac{\lambda_{min}(A^T A)}{\|A\|_F^2}, \ \beta = 0, \ for \ k \geq n.$$

Then,

$$\mathbb{E}\|x^{(k)} - x^*\|^2 \leq \tilde{\alpha}^{k-n+1}\mathbb{E}\|x^{(n-1)} - x^*\|^2$$
$$\leq \tilde{\alpha}^{k-n+1}\left(\alpha^{n-1}\|x^{(0)} - x^*\|^2 + \sum_{l=0}^{n-2}\alpha^l\mu\|b_{\mathcal{R}(A)^{\perp}} - z^{(n-2-l)}\|^2\right).$$

Thus, when k is large enough, the upper bound of the convergence factor of the PREKR method is less than or equal to that of the PREK method in [55].

2.8.2 Maximum-distance Extended Kaczmarz Method

In the following, we select the other three the probability criteria associated with residuals in [59] for the first component in the REK method and the orthogonal projections in the second component are in the given cyclic order. The three selection rules are as follows

$$(1) \frac{\frac{\left|r_{i_k}^{(k)}\right|^2}{\|a_{i_k}\|^2}}{\sum\limits_{i=1}^{m} \frac{\left|r_i^{(k)}\right|^2}{\|a_i\|^2}},$$

$$(2) \max_{1 \leqslant i \leqslant m} \frac{\left|b_i - \langle a_i, x^{(k)} \rangle - z_i^{(k)}\right|^2}{\|a_i\|^2},$$

$$(3) \max_{1 \leqslant i \leqslant m} \left|b_i - \langle a_i, x^{(k)} \rangle - z_i^{(k)}\right|.$$

Algorithm 2.8.2 Maximum-distance Extended Kaczmarz(MDEK) Algorithm

---

procedure $(A, b, l, x^{(0)} \in \mathcal{R}(A^T)\backslash\{0\}, z^{(0)} = b)$

    for $k = 0, 1, \cdots, l-1$ **do**

        Choose $i_k = arg \max\limits_{1 \leqslant i \leqslant m} \dfrac{\left|b_i - \langle a_i, x^{(k)} \rangle - z_i^{(k)}\right|^2}{\|a_i\|^2}$

        Set $x^{(k+1)} = x^{(k)} + \dfrac{b_{i_k} - z_{i_k}^{(k)} - \langle a_{i_k}, x^{(k)} \rangle}{\|a_{i_k}\|^2} a_{i_k}$

        Select $j_k = (k \bmod n)+1$

        Set $z^{(k+1)} = z^{(k)} - \dfrac{A_{j_k}^T z^{(k)}}{\|A_{j_k}\|^2} A_{j_k}$

    end for

    Output $x^{(l)}$.

end procedure

We call them the partially randomized Kaczmarz method with residual homogenizing (PREKRH), the maximum-distance extended Kaczmarz (MDEK) method, and the maximal-residual extended Kaczmarz (MREK) method, respectively. The MREK method presents in Algorithm 2.8.2, the other two methods can be analogously obtained. Now we give the following convergence theorems[56].

Theorem 2.8.2. *Let $A \in \mathbb{R}^{m \times n}$ be the coefficient matrix of the linear system, α_1, β_1 be defined as in (2.8.9). Then, the iteration sequence $\{x^{(k)}\}_{k=0}^{\infty}$ generated by the PREKRH method with $x^{(0)} \in \mathcal{R}(A^T) \backslash \{0\}$ converges to the least-squares solution $x_* = A^\dagger b$ in expectation. Moreover, the solution error in expectation for the iteration sequence $\{x^{(k)}\}_{k=0}^{\infty}$ satisfies*

$$E\left\|x^{(k)} - x^*\right\|^2 \leqslant \alpha^k \|x^{(0)} - x^*\|^2 + \frac{\left(\alpha^{k-\lfloor \frac{k}{2n} \rfloor \cdot n} + \beta^{\lfloor \frac{k}{2n} \rfloor}\right) \mu \gamma \kappa^2(A)}{\alpha_1^2} \|x^*\|^2,$$

where

$$\alpha = 1 - \frac{\alpha_1^2 \lambda_{min}(A^T A)}{\gamma}, \quad \beta = 1 - \frac{det(A^T A)}{\prod\limits_{j=1}^{n} \|A_j\|^2}, \quad \mu = \frac{\alpha_1 \beta_1}{\gamma} + \frac{1 + \beta_1}{\min\limits_{1 \leqslant i \leqslant m} \|a_i\|^2}$$

with $\gamma = (m-1) \max\limits_{1 \leqslant i \leqslant m} \|a_i\|^2.$

Theorem 2.8.3. *Assume that $A \in \mathbb{R}^{m \times n}$ is the coefficient matrix of the linear system, and α_1, β_1 are two real numbers defined as in (2.8.9). Then the iteration sequence $\{x^{(k)}\}_{k=0}^{\infty}$ generated by the MDEK method with $x^{(0)} \in \mathcal{R}(A^T) \backslash \{0\}$ converges to the least-squares solution $x^* = A^\dagger b$. Moreover, the solution error for the iteration sequence $\{x^{(k)}\}_{k=0}^{\infty}$ holds*

$$\left\|x^{(k)} - x^*\right\|^2 \leqslant \alpha^k \|x^{(0)} - x^*\|^2 + \frac{\left(\alpha^{k-\lfloor \frac{k}{2n} \rfloor \cdot n} + \beta^{\lfloor \frac{k}{2n} \rfloor}\right) \mu \gamma \kappa^2(A)}{\alpha_1^2} \|x^*\|^2,$$

$$(2.8.20)$$

where

$$\alpha = 1 - \frac{\alpha_1^2 \lambda_{min}(A^T A)}{\gamma}, \quad \beta = 1 - \frac{det(A^T A)}{\prod\limits_{j=1}^{n} \|A_j\|^2}, \quad \mu = \frac{\alpha_1 \beta_1}{\gamma} + \frac{1 + \beta_1}{\min\limits_{1 \leqslant i \leqslant m} \|a_i\|^2}$$

with $\gamma = \|A\|_F^2 - \min\limits_{1\leqslant i\leqslant m} \|a_i\|^2.$

Proof. Denoting $c_i^{(k)} = \dfrac{\left|r_i^{(k)}\right|^2}{\|a_i\|^2}, 1\leqslant i\leqslant m$, with the choice

$$i_k = arg\max\limits_{1\leqslant i\leqslant m} \frac{\left|b_i - \langle a_i, x^{(k)}\rangle - z_i^{(k)}\right|^2}{\|a_i\|^2},$$

we can get

$$
\begin{aligned}
\frac{\left|b_{i_k} - \langle a_{i_k}, x^{(k)}\rangle - z_{i_k}^{(k)}\right|^2}{\|a_{i_k}\|^2}
&= \frac{\max\limits_{1\leqslant i\leqslant m} \frac{\left|b_i - \langle a_i, x^{(k)}\rangle - z_i^{(k)}\right|^2}{\|a_i\|^2}}{\|r^{(k)}\|^2} \|r^{(k)}\|^2 \\
&= \frac{\max\limits_{1\leqslant i\leqslant m} c_i^{(k)}}{\sum\limits_{i=1}^{m} c_i^{(k)}\|a_i\|^2} \|r^{(k)}\|^2 \\
&= \frac{\max\limits_{1\leqslant i\leqslant m} c_i^{(k)}}{\sum\limits_{\substack{i=1 \\ i\neq i_{k-1}}}^{m} c_i^{(k)}\|a_i\|^2} \|r^{(k)}\|^2 \qquad (2.8.21) \\
&\geqslant \frac{\|r^{(k)}\|^2}{\sum\limits_{\substack{i=1 \\ i\neq i_{k-1}}}^{m} \|a_i\|^2} \\
&\geqslant \frac{\alpha_1\|A(x^{(k)} - x^*)\|^2 - \beta_1\|b_{\mathcal{R}(A)^\perp} - z^{(k)}\|^2}{\gamma}.
\end{aligned}
$$

The last inequality is obtained by Lemma 2.8.4, where $\gamma = \|A\|_F^2 - \min\limits_{1\leqslant i\leqslant m} \|a_i\|^2.$ Considering the process of the proof for Theorem 2.8.1, with this option for i_k we know that (2.8.16) becomes

$$
\begin{aligned}
&\|x^{(k+1)} - x^*\|^2 \\
&= \|x^{(k)} - x^*\|^2 - \frac{\left|\langle a_{i_k}, x^{(k)} - x^*\rangle\right|^2}{\|a_{i_k}\|^2} + \frac{\left|(b_{\mathcal{R}(A)^\perp})_{i_k} - z_{i_k}^{(k)}\right|^2}{\|a_{i_k}\|^2} \\
&= \|x^{(k)} - x^*\|^2 - \frac{\left|(b_{\mathcal{R}(A)})_{i_k} - \langle a_{i_k}, x^{(k)}\rangle\right|^2}{\|a_{i_k}\|^2} + \frac{\left|(b_{\mathcal{R}(A)^\perp})_{i_k} - z_{i_k}^{(k)}\right|^2}{\|a_{i_k}\|^2}
\end{aligned}
$$

$$\leqslant \; \|x^{(k)} - x^*\|^2 - \frac{\alpha_1 \left| b_{i_k} - \langle a_{i_k}, x^{(k)} \rangle - z_{i_k}^{(k)} \right|^2}{\|a_{i_k}\|^2} + (1+\beta_1) \frac{\left| (b_{\mathcal{R}(A)^\perp})_{i_k} - z_{i_k}^{(k)} \right|^2}{\|a_{i_k}\|^2}$$

$$\leqslant \; \|x^{(k)} - x^*\|^2 - \frac{\alpha_1^2 \|A(x^{(k)} - x^*)\|^2}{\gamma} + \frac{\alpha_1 \beta_1 \|b_{\mathcal{R}(A)^\perp} - z^{(k)}\|^2}{\gamma}$$

$$+ \frac{(1+\beta_1) \left| (b_{\mathcal{R}(A)^\perp})_{i_k} - z_{i_k}^{(k)} \right|^2}{\|a_{i_k}\|^2}$$

$$\leqslant \; \left(1 - \frac{\alpha_1^2 \lambda_{min}(A^T A)}{\gamma}\right) \|x^{(k)} - x^*\|^2$$

$$+ \left(\frac{\alpha_1 \beta_1}{\gamma} + \frac{1+\beta_1}{\min\limits_{1 \leqslant i \leqslant m} \|a_i\|^2}\right) \left\| b_{\mathcal{R}(A)^\perp} - z^{(k)} \right\|^2.$$

By the inequality (2.8.21) we obtained the second inequality. Since $x_k - x_* \in \mathcal{R}(A)$, we have $\|Az\|^2 \geqslant \lambda_{min}(A^T A)\|z\|^2$. Thus, the last inequality holds. Let $\alpha = 1 - \frac{\alpha_1^2 \lambda_{min}(A^T A)}{\gamma}$, $\mu = \frac{\alpha_1 \beta_1}{\gamma} + \frac{1+\beta_1}{\min\limits_{1 \leqslant i \leqslant m} \|a_i\|^2}$. With the analogous analysis in Theorem 2.8.1, the conclusions in Theorem 2.8.3 are true. $\qquad\square$

Theorem 2.8.4. *Suppose that $A \in \mathbb{R}^{m \times n}$ is the coefficient matrix of the linear system, α_1, β_1 are as in (2.8.9). Then the iteration sequence $\{x^{(k)}\}_{k=0}^{\infty}$ generated by the MREK method with $x^{(0)} \in \mathcal{R}(A^T) \backslash \{0\}$ converges to the least-squares solution $x^* = A^\dagger b$. Furthermore, the solution error for the iteration sequence $\{x^{(k)}\}_{k=0}^{\infty}$ satisfies*

$$\left\| x^{(k)} - x^* \right\|^2 \leqslant \alpha^k \|x^{(0)} - x^*\|^2 + \frac{\left(\alpha^{k - \lfloor \frac{k}{2n} \rfloor \cdot n} + \beta^{\lfloor \frac{k}{2n} \rfloor} \right) \mu \gamma \kappa^2(A)}{\alpha_1^2} \|x^*\|^2,$$

$$(2.8.22)$$

where

$$\alpha = 1 - \frac{\alpha_1^2 \lambda_{min}(A^T A)}{\gamma}, \quad \beta = 1 - \frac{det(A^T A)}{\prod\limits_{j=1}^{n} \|A_j\|^2}, \quad \mu = \frac{\alpha_1 \beta_1}{\gamma} + \frac{1+\beta_1}{\min\limits_{1 \leqslant i \leqslant m} \|a_i\|^2}$$

with $\quad \gamma = (m-1) \max\limits_{1 \leqslant i \leqslant m} \|a_i\|^2.$

Remark 2.8.2. For solving large inconsistent linear systems, the extended Kaczmarz methods with random sampling and maximum-distance are proposed in [60].

2.9 A Note: Minimum-Norm Solution of Least Square Problem

The least squares solution is not necessarily unique, and one of the ways to solve is relatively easy. A rich solution is that the least squares solution with the least norm or the minimum-norm solution of the least squares solution: x_{LNLS}.

$$x_{LNLS} = argmin\,\{\|x\| :\ argmin\|Ax - b\|\}. \qquad (2.9.1)$$

Proposition 2.9.1. *The least squares solutions of the systems of equations $Ax = b$, with $A \in \mathcal{R}^{m \times n}$ and $b \in \mathcal{R}^m$, can be represented as:*

$$x_{LS} = A^+ b + (I_n - A^+ A)y, \qquad (2.9.2)$$

where $y \in \mathcal{R}^n$ is an arbitrary vector and $I_n \in \mathcal{R}^{n \times n}$ is the identity matrix, and the least squares solution with the least norm x_{LNLS} is unique and $x_{LNLS} = A^+ b$.

Proof. In fact, set $rank(A) = r < n$, we have the full rank decomposition:

$$A_{m \times n} = F_{m \times r} G_{r \times n}, \qquad (2.9.3)$$

and we can check that the Moore-Penrose generalized inverse is

$$A^+ = G^T (GG^T)^{-1} (F^T F)^{-1} F^T. \qquad (2.9.4)$$

We know that η is the least squares solutions of $Ax = b$ if and only if η satisfies the normal equation

$$A^T A\eta = A^T b. \qquad (2.9.5)$$

By (2.9.3) and (2.9.4), it is easy to prove that (2.9.5) is equivalent to

$$A^+ Ax = A^+ b. \tag{2.9.6}$$

With the use of the expression of A^+, we can get $A^T AA^+ = A^T$, and

$$A^T Ax_{LS} = A^T AA^+ b + A^T Ay - A^T AA^+ Ay = A^T b + A^T Ay - A^T Ay = A^T b.$$

That is, x_{LS} in (2.9.2) is the least square solution of $Ax = b$.

It is easy to check that $A^+ b$ is a least square solution of $Ax = b$, and $A^+ A$ is a symmetric idempotent matrix ($(A^+ A)^T = A^+ A$, $(A^+ A)^2 = A^+ A$). For the idempotent matrix $A^+ A$, the relation of the **Null Space** of $A^+ A$ and the **Range** of $I_n - A^+ A$ is as follows:

$$\mathcal{N}(A^+ A) = \mathcal{R}(I_n - A^+ A). \tag{2.9.7}$$

Let $z = x_{LS} - A^+ b$, based on $A^+ Az = 0$, we know that $z \in \mathcal{N}(A^+ A)$. Therefore, $z \in \mathcal{R}(I_n - A^+ A)$, that is, $z = (I_n - A^+ A)y$, for some $y \in \mathcal{R}^n$. Thus, we get

$$x_{LS} = A^+ b + (I_n - A^+ A)y.$$

By $x_{LS} = A^+ b + z$ and $A^+ Az = 0$, we have

$$A^+ Az = G^T (GG^T)^{-1} Gz = 0. \tag{2.9.8}$$

The both sides of last equality of (2.9.8) are multiplied by G, which becomes

$$GG^T (GG^T)^{-1} Gz = 0,$$

that is, $Gz = 0$. Therefore,

$$(A^+ b)^T z = b^T F(F^T F)^{-1}(GG^T)^{-1} Gz = 0.$$

Thus,

$$\|x_{LS}\|^2 = \|A^+ b + z\|^2 = \|A^+ b\|^2 + \|z\|^2 + 2(A^+ b)^T z = \|A^+ b\|^2 + \|z\|^2 \geqslant \|A^+ b\|^2. \tag{2.9.9}$$

This proves that A^+b is a least squares solution with the least norm. If there is another least squares solution with the least norm x_1, then $\|x_1\| = \|A^+b\|$. Replace x_{LS} with x_1 in (2.9.9), we get $\|z\| = \|x_1\| - \|A^+b\| = 0$, and so $x_1 = A^+b$. The uniqueness is proved. $\qquad\square$

Remark 2.9.1. *MATLAB function* **LSMIN** *calculates the minimum-norm solution of the least squares problem $Ax = b$, where A is low-rank matrix. The function LSMIN is faster than the MATLAB alternative $x = pinv(A)*b$.*

Remark 2.9.2. **REK** *and* **REGS** *can also do it for any over- or under-determined systems.*

2.10   RK or RGS for Ridge Regression

For statistical as well as computational reasons, one often prefers to solve what is called ridge regression or Tikhonov-regularized least squares regression. This corresponds to solving the convex optimization problem[49]

$$\min_x \left\{ \|b - Ax\|^2 + \lambda\|x\|^2 \right\} \qquad (2.10.1)$$

for a given parameter λ (which we assume is fixed and known here) and a (real or complex) $m \times n$ matrix A and m-dimensional vector b.

There exist a large number of algorithms, iterative and not, randomized and not, for this problem. As for subclass of randomized algorithms, RK and RGS, the convergence rates of variants of RK and RGS for ridge regression posses linear convergence in expectation. The emphasis will be on the convergence rate parameters (e.g., condition number) that come into play for these algorithms when $m > n$ and $m < n$. It is showed that when $m > n$, one should randomize over columns (RGS), and when $m < n$, one should randomize over rows (RK).

RK and RGS as variants of randomized coordinate descent. Both RK and RGS can be viewed in the following fashion: suppose we have a positive definite matrix A, and we want to solve $Ax = b$. Casting the solution to the linear system as the solution to $\min_x \left\{ \frac{1}{2} x^T A x - x^T b \right\}$, one can derive the coordinate descent update

$$x^{(k+1)} = x^{(k)} + \frac{b_i - \langle a_i, x^{(k)} \rangle}{A_{ii}} e_i. \qquad (2.10.2)$$

where $\langle a_i, x^{(k)} \rangle - b_i$ is basically the ith coordinate of the gradient, and A_{ii} is the Lipschitz constant of the ith coordinate of the gradient (see related works, e.g., Leventhal and Lewis[20], Nesterov[61], Lee and Sidford[62]). In this light, the original RK update in (2.1.1) can be seen as the randomized coordinate descent rule for the positive semi-definite system $A A^T y = b$ (using the standard primal-dual mapping $x = A^T y$) and treating A as $A A^T$. Similarly, the RGS update in (2.5.1) can be seen as the randomized coordinate descent rule for the positive semi-definite system $A^T A x = A^T b$ and treating A as $A^T A$ and b as $A^T b$.

Variants of RK and RGS for ridge regression.   Utilizing the connection to coordinate descent, we can derive two algorithms for ridge regression, depending on how we formulate the linear system that solves the underlying optimization problem. In the first formulation, we let x^* be the solution of the system

$$\left(A^T A + \lambda I_n \right) x = A^T b, \qquad (2.10.3)$$

and using columns of A we attempt to solve for x^* iteratively by updating an initial guess $x^{(0)}$. In the second, we note the identity $x^* = A^T y^*$, where y^* is the optimal solution of the system

$$\left(A A^T + \lambda I_m \right) y = b, \qquad (2.10.4)$$

and using rows of A we attempt to solve for y^* iteratively by updating an initial guess $y^{(0)}$. The formulations (2.10.3) and (2.10.4) can be viewed as primal and dual variants, respectively, of the ridge regression problem.

It is a short exercise to verify that the optimal solutions of these two seemingly different formulations are actually identical (in the machine learning literature, the latter is simply an instance of kernel ridge regression). The second method's randomized coordinate descent updates are

$$\delta_i^{(k)} = \frac{b_i - a_i^T A^T y^{(k)} - \lambda y_i}{\|a_i\|^2 + \lambda}, \tag{2.10.5}$$

$$y_i^{(k+1)} = y_i^{(k)} + \delta_i^{(k)}, \tag{2.10.6}$$

where the *ith* row is selected with probability proportional to $\|a_i\|^2 + \lambda$. We may keep track of x as y changes with the update $x^{(k+1)} = x^{(k)} + \delta_i^{(k)} a_i$. Denote $K = AA^T$ as the Gram matrix of inner products between rows of A, and $r_i^{(k)} = b_i - \sum_j K_{ij} y_j^{(k)}$ as the *ith* residual at step k, the above update for y can be rewritten as

$$y_i^{(k+1)} = \frac{K_{ii}}{K_{ii} + \lambda} y_i^{(k)} + \frac{b_i - \sum_j K_{ij} y_j^{(k)}}{K_{ii} + \lambda} = S_{\frac{\lambda}{K_{ii}}} \left(y_i^{(k)} + \frac{r_i^{(k)}}{K_{ii}} \right), \tag{2.10.7}$$

where row i is picked with probability proportional to $K_{ii} + \lambda$ and $S_\mu(z) := \frac{z}{1+\mu}$.

In contrast, we write below the randomized coordinate descent updates for the first linear system. Analogously calling $r^{(k)} = b - Ax^{(k)}$ the residual vector, we have

$$x_j^{(k+1)} = x_j^{(k)} + \frac{A_j^T b - A_j^T Ax^{(k)} - \lambda x_j^{(k)}}{\|A_j\|^2 + \lambda} = S_{\frac{\lambda}{\|A_j\|^2}} \left(x_j^{(k)} + \frac{A_j^T r^{(k)}}{\|A_j\|^2} \right), \tag{2.10.8}$$

Computation and convergence. The algorithms presented here are of computational interest because they completely avoid inverting, storing, or even forming AA^T and $A^T A$. The RGS ((2.10.8) working on columns) updates take $O(m)$ time, since each column (feature) is of size m. In contrast, the RK ((2.10.7) working on rows) updates take $O(n)$ time since that is the length of a row (data point). While the RK and RGS algorithms are similar and related, one should not be tempted into thinking their convergence rates are the same. Indeed, using a similar style proof as presented in [46], one can analyze the convergence rates in parallel as follows. Let us denote

$$\Sigma' := A^T A + \lambda I_n \in \mathcal{R}^{n \times n}, \quad and \quad K' := AA^T + \lambda I_m \in \mathcal{R}^{m \times m}$$

for brevity, and let $\sigma_1, \sigma_2, \cdots$ be the singular values of A in increasing order. Observe that

$$\sigma_{min}(\Sigma') = \begin{cases} \sigma_1^2 + \lambda, & if \ m \geqslant n \\ \lambda, & if \ m < n \end{cases} \quad and \ \sigma_{min}(K') = \begin{cases} \lambda, & if \ m > n \\ \sigma_1^2 + \lambda, & if \ m \leqslant n. \end{cases}$$

Then, denoting x^* and y^* as the solutions to the two ridge regression formulations and $x^{(0)}$ and $y^{(0)}$ as the initializations of the two algorithms, the following result is proved.

Theorem 2.10.1. *The rate of convergence for RK for ridge regression is*

$$E\|y^{(k)} - y^*\|_{K+\lambda I_m} \leqslant \begin{cases} \left(1 - \dfrac{\lambda}{\sum\limits_i \sigma_i^2 + m\lambda}\right)^k \|y^{(0)} - y^*\|_{K+\lambda I_m}^2, & if \ m > n, \\[4mm] \left(1 - \dfrac{\sigma_1^2 + \lambda}{\sum\limits_i \sigma_i^2 + m\lambda}\right)^k \|y^{(0)} - y^*\|_{K+\lambda I_m}^2, & if \ m \leqslant n. \end{cases}$$

$$(2.10.9)$$

The rate of convergence for RGS for ridge regression is

$$E\|x^{(k)} - x^*\|_{A^T A + \lambda I_n} \leqslant \begin{cases} \left(1 - \dfrac{\sigma_1^2 + \lambda}{\sum\limits_i \sigma_i^2 + n\lambda}\right)^k \|x^{(0)} - x^*\|_{A^T A + \lambda I_n}^2, & if \ m \geqslant n, \\[4mm] \left(1 - \dfrac{\lambda}{\sum\limits_i \sigma_i^2 + n\lambda}\right)^k \|x^{(0)} - x^*\|_{A^T A + \lambda I_n}^2, & if \ m < n. \end{cases}$$

$$(2.10.10)$$

Immediately, it is noted that RGS is preferable in the over-determined case while RK is preferable in the under-determined case. Hence, the proposal for solving such systems is as follows:

When $m > n$, RGS is always used, and when $m < n$, RK is always used.

Remark 2.10.1. *The two methods' convergence analyses are based on the full rank (full column rank for RK and full row rank for RGS), that is, the* two methods are to solve the unique least square solution or the unique least l_2 norm solution. How about the case of rank deficient? or, how to solve x_{LNLS}?

Remark 2.10.2. One can view x_{RR} and y_{RR} simply as solutions to the two linear systems

$$(A^T A + \lambda I_n)x = A^T b, \quad and \quad (AA^T + \lambda I_m)y = b.$$

If we naively use RK or RGS on either of these systems (treating them as solving $Ax = b$ for some given A and b), we may apply the bounds (2.1.3) and (2.5.4) to the matrix $A^T A + \lambda I_n$ or $AA^T + \lambda I_m$. This, however, yields a bound on the convergence rate which depends on the squared scaled condition number of $A^T A + \lambda I_n$, which is approximately the fourth power of the scaled condition number of A. This dependence is suboptimal, so that using these methods it becomes highly impractical to solve large-scale problems. This is of course not surprising since this naive solution does not utilize any structure of the ridge regression problem (for example, the aforementioned matrices are positive definite). One thus searches for more tailored approaches-- indeed. Here it is proposed RK and RGS updates whose computation are still only $O(n)$ or $O(m)$ per iteration and yield linear convergence with dependence only on the scaled condition number of $A^T A + \lambda I_n$ or $AA^T + \lambda I_m$, and not their square. The aforementioned updates and their convergence rates are motivated by a clear understanding of how RK and RGS methods related to each other as in [46] and jointly to positive semidefinite systems of equations.

Is there an algorithm that suitably randomizes over both rows and columns? For example:

1: Intialize $x^{(0)} = 0$, $y^{(0)} = 0$ and $z^{(0)} = b$

2: for $k = 0, 1, \cdots$ do

3:     Pick $i_k \in [m]$ with probability $q_{i_k} = \|a_{i_k}^T\|^2 / \|A\|_F^2$

4:     Pick $j_k \in [n]$ with probability $q_{j_k} = \|A_{j_k}\|^2 / \|A\|_F^2$

5:     Set $z^{(k+1)} = z^{(k)} - \dfrac{\langle A_{j_k}, z^{(k)} \rangle}{\|A_{j_k}\|^2} A_{j_k}$

6:     Set $x^{(k+1)} = x^{(k)} - \dfrac{b_{i_k} - z_{i_k}^{(k)} - \langle a_{i_k}, x^{(k)} \rangle}{\|a_{i_k}\|^2} a_{i_k}$

7:    *Set $P_{i_k} = I_n - \frac{a_{i_k} a_{i_k}}{\|a_{i_k}\|^2}$,*

8:    *Set $y^{(k+1)} = P_{i_k}(y^{(k)} + x^{(k+1)} - x^{(k)})$,*

9:    *Check $x_{LNLS} = x^{(k+1)} - y^{(k+1)}$ every $8\min(m, n)$ iterations and terminate if it holds:*

$$\frac{\|Ax_{LNLS} - (b - z^{(k+1)})\|}{\|A\|_F \|x_{LNLS}\|} \leqslant \epsilon \; , \quad \frac{\|Ay^{(k+1)}\|}{\|A\|_F \|x_{LNLS}\|} \leqslant \epsilon \;\; and \;\; \frac{\|A^T z^{(k+1)}\|}{\|A\|_F \|x_{LNLS}\|} \leqslant \epsilon.$$

10:   *end for*

11:   *Output x_{LNLS}.*

2.11   Randomized Block Kaczmarz Method

Randomized iterative methods have attracted much attention in recent years because they can approximately solve large-scale linear systems of equations without accessing the entire coefficient matrix. Block variants of the RK and RCD (RGS) methods have been proposed to accelerate the convergence (see, e.g.,[63–67]). Although block random selection does not necessarily provide the best order, like the single row random Kaczmarz method, this algorithm can provide a linear convergence rate.

Algorithm 2.11.1 Randomized Block Kaczmarz (RBK) Algorithm

---

procedure $(A, \; b, \; x^{(0)} \in \mathcal{R}(A^T))$

Set a partition $T = \{\tau_1, \tau_2, \cdots, \tau_p\}$ of the row indices $\{1, \cdots, m\}$

for $k = 0, 1, \cdots, K$ **do**

Select a block τ_k uniformly at random from T and set

$$x^{(k+1)} = x^{(k)} + A_{\tau_k}^+ \left(b_{\tau_k} - A_{\tau_k} x^{(k)} \right) \tag{2.11.1}$$

end for

Output: $x^{(K+1)}$

end procedure

The most expensive (arithmetic) step in Algorithm 2.11.1 occurs when we apply the pseudoinverse $A_{\tau_k}^+$ to a vector. We can perform this calculation efficiently provided that each submatrix A_{τ_k} has well conditioned rows. Indeed, in this case, we can invoke an iterative least-squares solver, such as CGLS, to apply the pseudoinverse $A_{\tau_k}^+$ approximately with the use of a small number of matrix-vector multiplies with A_{τ_k} and $A_{\tau_k}^T$. In particular, we never need to form the pseudoinverse[63].

Definition 2.11.1. [63] *(Row paving). An (p, α, β) row paving of a matrix A is a partition $T = \{\tau_1, \tau_2, \cdots, \tau_p\}$ of the row indices that verifies*

$$\alpha \leqslant \lambda_{\min}(A_{\tau_k} A_{\tau_k}^T) \quad and \quad \lambda_{\max}(A_{\tau_k} A_{\tau_k}^T) \leqslant \beta \quad for\ each\ \tau_k \in T.$$

The number p of blocks is called the size of the paving. The numbers α and β are called lower and upper paving bounds. The ratio β/α gives a uniform bound on the squared condition number $\kappa^2(A_{\tau_k})$ for each τ_k.

Note that $\alpha = 0$ unless each submatrix A_{τ_k} is fat. Every partition T of the rows of a matrix A has associated paving parameters (p, α, β). In a moment, we will see how these quantities play a role in the performance of the algorithm. Roughly speaking, it is best that the size p, the upper bound β, and the conditioning β/α of the paving are small.

Theorem 2.11.1. *(Convergence). Suppose A is a matrix with full column rank that admits an (p, α, β) row paving T (**Without losing generality, assume that A is standardized**). Consider the least-squares problem*

$$minimize\ \|Ax - b\|^2.$$

Let x^* be the unique minimizer, and define the residual $e := Ax^* - b$. For any initial estimate $x^{(0)}$, the randomized block Kaczmarz method (Algorithm 2.11.1) produces a sequence $\{x^{(j)} : j \geqslant 0\}$ of iterates that satisfies

$$E\|x^{(k+1)} - x^*\|^2 \leqslant \left[1 - \frac{\sigma_{\min}^2(A)}{\beta p}\right]^k \|x^{(0)} - x^*\|^2 + \frac{\beta}{\alpha} \cdot \frac{\|e\|^2}{\sigma_{\min}^2(A)}. \quad (2.11.2)$$

Proof. Based on the hypotheses and notation of the theorem, we firstly prove that

$$\|x^{(k+1)} - x^*\|^2 \leqslant \| \left(I - A_{T_k}^+ A_{T_k}\right) (x^{(k)} - x^*)\|^2 + \frac{1}{\alpha}\|e_{T_k}\|^2. \qquad (2.11.3)$$

In fact, according to the update rule (2.11.1), block Kaczmarz computes

$$x^{(k+1)} = x^{(k)} + A_{T_k}^+ \left(b_{T_k} - A_{T_k}x^{(k)}\right) = x^{(k)} + A_{T_k}^+ A_{T_k}(x^* - x^{(k)}) - A_{T_k}^+ e_{T_k},$$

where we have introduced the decomposition $b = Ax^* - e$, restricted to the coordinates listed in T_k. Subtract x^* from both sides to obtain

$$x^{(k+1)} - x^* = \left(I - A_{T_k}^+ A_{T_k}\right)(x^{(k)} - x^*) - A_{T_k}^+ e_{T_k}$$

The range of $A_{T_k}^+$ and the range of $I - A_{T_k}^+ A_{T_k}$ are orthogonal, so we may invoke the Pythagorean Theorem to reach

$$\|x^{(k+1)} - x^*\|^2 = \| \left(I - A_{T_k}^+ A_{T_k}\right)(x^{(k)} - x^*)\|^2 + \|A_{T_k}^+ e_{T_k}\|^2$$

The second term on the right-hand side satisfies

$$\|A_{T_k}^+ e_{T_k}\|^2 \leqslant \sigma_{\max}^2(A_{T_k}^+)\|e_{T_k}\|^2 \leqslant \frac{\|e_{T_k}\|^2}{\sigma_{min}^2(A_{T_k})} \leqslant \frac{\|e_{T_k}\|^2}{\alpha},$$

where α is the lower bound on the row paving T. Combine the last two displays to wrap up.

Suppose that T_k is chosen uniformly at random from the row paving T. For fixed vectors u and v, it holds that

$$E\| \left(I - A_{T_k}^+ A_{T_k}\right) u\|^2 \leqslant \left[1 - \frac{\sigma_{\min}^2(A)}{\beta p}\right] \|u\|^2 \quad and \quad E\|v_{T_k}\|^2 = \frac{\|v\|^2}{p}.$$

$$(2.11.4)$$

In fact, the second identity of (2.11.4) emerges from a very short calculation:

$$E\|v_{T_k}\|^2 = \frac{\sum_{T_i \in T} \|v_{T_i}\|^2}{p} = \frac{\|v\|^2}{p},$$

which depends on the fact that the blocks of T partition the components of v.

Since $A_{\tau_k}^+ A_{\tau_k}$ is an orthogonal projector, we may apply the Pythagorean Theorem to obtain the relation

$$E\| \left(I - A_{\tau_k}^+ A_{\tau_k}\right) u \|^2 = \|u\|^2 - E\|A_{\tau_k}^+ A_{\tau_k} u\|^2.$$

We control the remaining expectation as follows.

$$E\|A_{\tau_k}^+ A_{\tau_k} u\|^2 \geqslant E[\sigma_{min}^2(A_{\tau_k}^+)\|A_{\tau_k} u\|^2] \geqslant \frac{1}{\beta} \cdot E\|A_{\tau_k} u\|^2$$

$$= \frac{1}{\beta p} \sum_{\tau_i \in T} \|A_{\tau_i} u\|^2 = \frac{1}{\beta p}\|Au\|^2 \geqslant \frac{\sigma_{min}^2(A)}{\beta p}\|u\|^2.$$

The second inequality depends on the bound $\sigma_{min}^2(A_{\tau_k}^+) = \sigma_{max}^{-2}(A_{\tau_k}) \geqslant \beta^{-1}$. The fourth relation holds because the blocks in a paving partition the row indices of A. To complete the proof of (2.11.4), we simply combine the last two displays.

The main result follows quickly once from (2.11.3) and (2.11.4).

First, we bound the expected error at iteration $k + 1$ in terms of the error at iteration k. Average the bound from (2.11.3) over the randomness in τ_k to reach

$$E_k\|x^{(k+1)} - x^*\|^2 \leqslant E_k\| \left(I - A_{\tau_k}^+ A_{\tau_k}\right)(x^{(k)} - x^*)\|^2 + \frac{1}{\alpha}E_k\|e_{\tau_k}\|^2.$$

$$\leqslant \left[1 - \frac{\sigma_{min}^2(A)}{\beta p}\right]\|x^{(k+1)} - x^*\|^2 + \frac{\|e\|^2}{\alpha p}.$$

The second inequality follows from (2.11.4), with $u = x^{(k)} - x^*$. and $v = e$.

By applying this result repeatedly, we can control the expected error after k iterations in terms of the initial error. Abbreviating $\gamma := 1 - \sigma_{min}^2(A)/(\beta p)$, we obtain the estimate

$$E\|x^{(k)} - x^*\|^2 = E_{\tau_1} E_{\tau_2} \cdots E_{\tau_k}\|x^{(k)} - x^*\|^2$$
$$\leqslant \gamma^j\|x^{(0)} - x^*\|^2 + \frac{1}{\alpha p}\|e\|^2 \left(\Sigma_{i=0}^{k-1}\gamma^i\right)$$
$$\leqslant \gamma^j\|x^{(0)} - x^*\|^2 + \frac{1}{\alpha p(1-\gamma)}\|e\|^2.$$

Reintroduce the value of γ in this expression, and simplify to complete the proof. $\qquad\square$

The expression (2.11.2) states that the block Kaczmarz method exhibits an expected linear rate of convergence until it reaches a ball about the true solution. The radius of this ball, which we call the convergence horizon, is comparable with the second term on the right-hand side of (2.11.2). The bracket controls the convergence rate. The minimum singular value of A affects both the rate of convergence and the convergence horizon. In each case, we prefer $\sigma_{\min}(A)$ to be as large as possible.

The properties of the row paving play an interesting role in Theorem 2.11.1. Curiously, the rate of convergence depends only on the upper paving bound β and the number of blocks in the paving. On the other hand, the convergence horizon reflects the conditioning β/α of the paving. Thus, the conditioning of the paving only affects the error bound when the least-squares problem is inconsistent (i.e., e is nonzero).

2.12　Randomized Block CD Method

Assume that A is full column rank ($m \geqslant n$). In the RGS (RCD) method, each iteration greedily minimizes the objective function

$$\min_{x \in \mathcal{R}^n} L(x) := \min_{x \in \mathcal{R}^n} \frac{1}{2} \|b - Ax\|^2, \quad A \in \mathcal{R}^{m \times n},\ b \in \mathcal{R}^m, \qquad (2.12.1)$$

with respect to a selected coordinate with probability proportional to corresponding column norms of A, and the results converge to the least-squares solution at an expected linear rate. Inspired by the block Kaczmarz algorithm,[67] demonstrated that the block method can also be applied to the RGS method to increase the convergence rate.

Let x^* be the unique minimizer of (2.12.1). Define column paving of $A = (A_{\tau_1}, A_{\tau_2}, \cdots, A_{\tau_s})$ as (s, α, β): $P = \{\tau_1, \tau_2, \cdots, \tau_s\}$ such that

$$\begin{cases} \alpha \geqslant \lambda_{\max}\left(A_{\tau_k}^T A_{\tau_k}\right) = \sigma_{\max}^2(A_{\tau_k}), \\ \sigma_{min}^2(A_{\tau_k}) = \lambda_{\min}\left(A_{\tau_k}^T A_{\tau_k}\right) \geqslant \beta, \end{cases} \quad for\ all\ \tau_k \in P. \qquad (2.12.2)$$

Let $I = (I_{\tau_1}, I_{\tau_2}, \cdots, I_{\tau_p})$ be partitions of the $n \times n$ identity matrix I corresponding to the column index of matrix A.

The algorithm for the Randomized Block Coordinate-Descent (RBCD) method or Randomized Block Gauss-Seidel (RBGS) method is described by the following code. Similar to the RGS method, the RBGS Method iteratively improves the approximation by adjusting the value of multiple coordinates, which finally converges to the least square solution x^*.

Algorithm 2.12.1 Randomized Block Gauss-Seidel (RBGS) Algorithm

---

procedure $(A, b, x^{(0)})$

Set a partition $P = \{\tau_1, \tau_2, \cdots, \tau_p\}$ of the column indices $\{1, \cdots, n\}$

for $k = 0, 1, \cdots, K$ **do**

Select a τ_k uniformly at random from P and compute

$$x^{(k+1)} = x^{(k)} + I_{\tau_i} A_{\tau_i}^+ r^{(k)}, \quad r^{(k+1)} = r^{(k)} - A_{\tau_i} A_{\tau_i}^+ r^{(k)} \qquad (2.12.3)$$

end for

Output: $x^{(K+1)}$

end procedure

Remark 2.12.1. *The iterative formula (2.12.3) can be written by*

$$w^{(k)} = A_{\tau_i}^+ r^{(k)}, \quad x_{\tau_i}^{(k+1)} = x_{\tau_i}^{(k)} + w^{(k)}, \quad r^{(k+1)} = r^{(k)} - A_{\tau_i} w^{(k)}. \qquad (2.12.4)$$

Theorem 2.12.1. *Let $A \in \mathcal{R}^{m \times n}$ and $b \in \mathcal{R}^m$. Assume the column of A partition (s, α, β) is defined as (2.12.2). For any initial estimate $x^{(0)} \in \mathcal{R}^n$, the Randomized Block Gauss-Seidel (RBGS) Method described in Algorithm 2.10.1 produces a sequence $\{x^{(k)}\}$ of iterates that satisfies:*

$$E\|x^{(k)} - x^*\|^2 \leqslant \kappa^2(A)\gamma^k \|x^{(0)} - x^*\|^2, \qquad (2.12.5)$$

where $\gamma = 1 - \frac{\sigma_{\min}^2(A)}{s\alpha}$ and $\kappa(A) = \|A\|_2 \cdot \|A^+\|_2$ is the condition number.

Proof. Firstly, it is easy to verify that $A(x^{(k+1)} - x^{(k)})$ and $A(x^{(k+1)} - x^*)$ are orthogonal. In fact, the conclusion is based on that $\mathcal{R}(A_{\tau_i}) \perp N(A_{\tau_i}^T)$,

$$A(x^{(k+1)} - x^{(k)}) = AI_{\tau_i} A_{\tau_i}^+ r^{(k)} = A_{\tau_i} A_{\tau_i}^+ r^{(k)} \in \mathcal{R}(A_{\tau_i}),$$

and

$$
\begin{aligned}
A_{\tau_i}^T A(x^{(k+1)} - x^*) &= I_{\tau_i}^T A^T A(x^{(k+1)} - x^*) = I_{\tau_i}^T A^T (Ax^{(k+1)} - b) \\
&= -A_{\tau_i}^T r^{(k+1)} = 0,
\end{aligned}
$$

so it follows that $A(x^{(k+1)} - x^*) \in N(A_{\tau_i}^T)$. The last equality is due to τ_k-block coordinate-descent.

Secondly, for any vector u,

$$
\begin{aligned}
\|A_{\tau_k} A_{\tau_k}^+ u\|^2 &= u^T A_{\tau_k} \left(A_{\tau_k}^T A_{\tau_k}\right)^{-1} A_{\tau_k}^T u \\
&= v^T \left(A_{\tau_k}^T A_{\tau_k}\right)^{-1} v \geqslant \lambda_{\max}^{-1}(A_{\tau_k}^T A_{\tau_k})\|v\|^2 \geqslant \frac{\|v\|^2}{\alpha},
\end{aligned}
$$

where $v - A_{\tau_k}^T u$. Take the expectation on both sides of the above, we get

$$
\begin{aligned}
E\|A_{\tau_k} A_{\tau_k}^+ u\|^2 &\geqslant \tfrac{1}{\alpha} E\|A_{\tau_k}^T u\|^2 \\
&= \tfrac{1}{s\alpha} \sum_{k=1}^{s} \|A_{\tau_k}^T u\|^2 \geqslant \tfrac{1}{s\alpha}\|A^T u\|^2 \geqslant \frac{\sigma_{\min}^2(A)}{s\alpha}\|u\|^2.
\end{aligned}
$$

Thirdly, if set $r = b - Ax^*$, then we have $A^T r = 0$ and $A_{\tau_k}^T r = 0$, thus

$$
A_{\tau_k}^+ r = \left(A_{\tau_k}^T A_{\tau_k}\right)^{-1} A_{\tau_k}^T r = 0.
$$

Now we give the proof of the theorem. By

$$
\|Ax^{(k+1)} - Ax^*\|^2 = \|Ax^{(k)} - Ax^*\|^2 - \|Ax^{(k+1)} - Ax^{(k)}\|^2,
$$

we have

$$
\begin{aligned}
&E\|Ax^{(k+1)} - Ax^*\|^2 \\
=\ & \|Ax^{(k)} - Ax^*\|^2 - E\|Ax^{(k+1)} - Ax^{(k)}\|^2 \\
=\ & \|Ax^{(k)} - Ax^*\|^2 - E\|A_{\tau_k} A_{\tau_k}^+ r^{(k)}\|^2 \\
=\ & \|Ax^{(k)} - Ax^*\|^2 - E\|A_{\tau_k} \left(A_{\tau_k}^T A_{\tau_k}\right)^{-1} I_{\tau_k}^T A^T r^{(k)}\|^2 \\
=\ & \|Ax^{(k)} - Ax^*\|^2 - E\|A_{\tau_k} A_{\tau_k}^+ A(x^* - x^{(k)})\|^2 \\
\leqslant\ & \|Ax^{(k)} - Ax^*\|^2 - \frac{\sigma_{\min}^2(A)}{s\alpha}\|A(x^* - x^{(k)})\|^2 \\
=\ & \left(1 - \frac{\sigma_{\min}^2(A)}{s\alpha}\right)\|A(x^* - x^{(k)})\|^2.
\end{aligned}
$$

Because A is full column rank, so $A^+ A = \left(A^T A\right)^{-1} A^T A = I_n$. Thus

$$
\|x^{(k+1)} - x^*\|^2 = \|A^+ A(x^{(k+1)} - x^*)\|^2 \leqslant \|A^+\|^2 \|Ax^{(k+1)} - Ax^*\|^2.
$$

Therefore,

$$
\begin{aligned}
E\|x^{(k+1)} - x^*\|^2 &\leqslant \|A^+\|^2 E\|Ax^{(k+1)} - Ax^*\|^2 \\
&\leqslant \|A^+\|^2 \left(1 - \frac{\sigma_{\min}^2(A)}{s\alpha}\right) \|A(x^* - x^{(k)})\|^2 \\
&\leqslant \|A^+\|^2 \|A\|^2 \left(1 - \frac{\sigma_{\min}^2(A)}{s\alpha}\right) \|x^* - x^{(k)}\|^2.
\end{aligned}
$$

From this, we can easily deduce the following inequalities

$$
E\|x^{(k)} - x^*\|^2 \leqslant \kappa^2(A) \left(1 - \frac{\sigma_{\min}^2(A)}{s\alpha}\right)^k \|x^* - x^{(0)}\|^2.
$$

$\square$

Similar to the proof of Theorem 2.12.1, we can get the following conclusion.

Theorem 2.12.2. [64] *Under the conditions of Theorem 2.12.1, the following conclusion holds*

$$
E\|A(x^{(k)} - x^*)\|^2 \leqslant \left(1 - \frac{\sigma_{\min}^2(A)}{s\alpha}\right)^k \|b_{\mathcal{R}(A)}\|^2.
$$

Remark 2.12.2. *If $s = 1$, then $\alpha = \|A\|^2$, and by Theorem 2.12.1 we get*

$$
E\|x^{(k)} - x^*\|^2 \leqslant \kappa^2(A) \left(1 - \frac{\sigma_{\min}^2(A)}{\sigma_{\max}^2(A)}\right)^k \|x^* - x^{(0)}\|^2.
$$

2.13 Randomized Double Block Kaczmarz Method

The bound (2.11.2) demonstrates that the RBK method performs well when the noise in inconsistent systems is small. Zouzias and Freris introduced a variant of the method which utilizes a random projection to iteratively reduce the norm of the error[40]. They show that the estimate of the REK method converges linearly in expectation to the least squares solution of the system, breaking the radius barrier of the standard method. The algorithm maintains not only an estimate $x^{(k)}$ to the solution but also an approximation $z^{(k)}$ to the projection of b onto the range of A. Now we consider a randomized double block Kaczmarz (RDBK) method to solve the least square solution of inconsistent systems of equations.

2.13.1   RDBK with Pseudoinverse

The RDBK method uses the RBK method which incorporates a blocked projection step, which provides accelerated convergence to the least squares solution. In this case it needs a column partition for the projection step and a row partition for the Kaczmarz step (randomized double block Kaczmarz) method. The results show that this method offers both linear convergence to the least squares solution x^* when a row paving can be obtained, and improved convergence speed due to the blocking of both the rows and columns.

Combining the theoretical approaches in [40, 63] it can be proved that the following result about the convergence of Algorithm 2.13.1. This result utilizes both a column paving and row paving.

Algorithm 2.13.1 RDBK Algorithm with Pseudoinverse

---

   procedure $(A,\ b,\ x^{(0)} = 0,\ K)$

       Set column partition $\mathcal{P} = \{v_1, \cdots, v_p\}$, row partition $\mathcal{Q} = \{\tau_1, \cdots, \tau_q\}$

       for $k = 1, 2, \cdots, K$ **do**

           Pick $\tau_k \in \mathcal{P}$ and $v_k \in \mathcal{Q}$ uniformly at random

           Set $z^{(k)} = z^{(k-1)} - A_{\tau_k}(A_{\tau_k})^+ z^{(k-1)}$

           Update $x^{(k)} = x^{(k-1)} + A_{v_k}^+ (b_{v_k} - z_{v_k}^{(k)} - A_{v_k} x^{(k-1)})$

       end for

       Output: $x^{(K+1)}$

   end procedure

Theorem 2.13.1. *Algorithm 2.13.1 with input $A \in \mathcal{R}^{m \times n}$, $b \in \mathcal{R}^m$, $T \in N$, $(p, \bar{\alpha}, \bar{\beta})$ column paving of A, and (q, α, β) row paving A, outputs an estimate vector $x^{(K)}$ that satisfies*

$$E\|X^{(K)} - x^*\|^2 \leqslant \gamma^K \|x^{(0)} - x^*\|^2 + \left(\gamma^{\lfloor K/2 \rfloor} + \bar{\gamma}^{\lfloor K/2 \rfloor}\right) \frac{\|b_{\mathcal{R}(A)}\|^2}{\alpha(1 - \gamma)},$$

where $\gamma = 1 - \frac{\sigma_{\min}^2(A)}{q\beta}$ *and* $\bar{\gamma} = 1 - \frac{\sigma_{\min}^2(A)}{p\beta}$.

2.13.2   Pseudoinverse-Free RDBK

Kui Du and Xiao-Hui Sun[68] proposed two novel pseudoinverse-free randomized block iterative algorithms for solving consistent and inconsistent linear systems: the block row uniform sampling (BRUS) algorithm and the block column uniform sampling (BCUS) algorithm. The BRUS and BCUS algorithms require two user-defined random matrices: one for row sampling and the other for column sampling. The linear convergence in the mean square sense of these two algorithms is proved.

Main theoretical results. For arbitrary initial guess $x^{(0)} \in \mathcal{R}^n$, define the vector

$$x_*^0 = A^+ b + (I - A^+ A) x^{(0)},$$

which is a solution if $Ax = b$ is consistent, or a least squares solution if $Ax = b$ is inconsistent. We mention that $x_*^{(0)}$ is the orthogonal projection of $x^{(0)}$ onto the set

$$\{x \in \mathcal{R}^n \mid A^T A x = A^T b\}.$$

The main theoretical results of this subsetion are as follows.

(1) The block row sampling iterative algorithm (one special case of the doubly stochastic block iterative algorithm) converges linearly in the mean square sense to x_*^0 if $Ax = b$ is consistent (Theorem 2.13.2) and to within a radius of x_*^0 if $Ax = b$ is inconsistent (Theorem 2.13.3).

(2) The block column sampling iterative algorithm (another special case of the doubly stochastic block iterative algorithm) converges linearly in the mean square sense to $A^+ b$ if A has full column rank (Theorem 2.13.4).

(3) The extended block row sampling iterative algorithm converges linearly in the mean square sense to x_*^0 for arbitrary linear system $Ax = b$ (there is no assumptions about the dimensions or rank of the coefficient matrix A and the system can be consistent or inconsistent (see Theorem 2.13.5)).

Lemma 2.13.1. *Let* $\alpha \geqslant 0$, $\beta \geqslant 0$, *and* $A \in \mathcal{R}^{m \times n}$ *be any nonzero matrix with* $\mathrm{rank}(A) = r$. *For all* $u \in \mathrm{range}(A^T)$ *and* $0 \leqslant i \leqslant k$, *it holds*

$$\|(I - \beta A^T A)^i (I - \alpha A^T A)^{k-i} u\| \leqslant \delta^k \|u\|,$$

where $\delta = \max\limits_{1 \leqslant i \leqslant r} \{|1 - \alpha \sigma_i^2(A)|, \ |1 - \beta \sigma_i^2(A)|\}$.

The proof is straightforward.

Article [68] first proposes a doubly stochastic block iterative (DSBI) algorithm, which is defined as follows:

$$x^{(k)} = x^{(k-1)} - \alpha T T^T A^T S S^T (A x^{(k-1)} - b), \tag{2.13.1}$$

where initial guess $x^{(0)} \in \mathcal{R}^n$, the stepsize parameter $\alpha > 0$, and the random parameter matrix pair $(S; T)$ are sampled independently in each iteration from a distribution $\mathcal{D}$ and satisfy

$$E[T T^T A^T S S^T] = A^T,$$

where $S \in \mathcal{R}^{m \times p}$ and $T \in \mathcal{R}^{n \times q}$, p and q are variable integers (and hence p and q are random variables).

The DSBI algorithm extends the DSGS algorithm proposed in [69]. In the followings, the case $T = I$ and the case $S = I$ are discussed, respectively.

When $T = I$, we refer to the DSBI algorithm as the block row sampling iterative (BRSI) algorithm. Given an arbitrary initial guess $x^{(0)} \in \mathcal{R}^n$, the kth iterate of the BRSI algorithm is

$$x^{(k)} = x^{(k-1)} - \alpha_r A^T S S^T (A x^{(k-1)} - b), \tag{2.13.2}$$

where the stepsize parameter $\alpha_r > 0$, and the random parameter matrix S is sampled independently in each iteration from a distribution $\mathcal{D}_r$ and satisfies $E[S S^T] = I$.

Define

$$\lambda_{\max}^r = \max_{S \sim \mathcal{D}_r} \lambda_{\max}(A^T S S^T A).$$

Theorem 2.13.2. *Assume that $0 < \alpha_r < 2/\lambda_{max}^r$. If $Ax = b$ is consistent, then for arbitrary $x^{(0)} \in \mathcal{R}^n$, the kth iterate $x^{(k)}$ of the BRSI algorithm satisfies*

$$E[\|x^{(k)} - x_*^0\|^2] \leqslant \eta_r^k \|x^{(0)} - x_*^0\|^2,$$

where $\eta_r = 1 - \alpha_r(2 - \alpha_r\lambda_{max}^r)\sigma_{min}^2(A)$.

Proof. It follows from $Ax_*^0 = b$ and

$$x^{(k)} - x_*^0 = x^{(k-1)} - x_*^0 - \alpha_r A^T SS^T(Ax^{(k-1)} - b) \tag{2.13.3}$$

that

$$\|x^{(k)} - x_*^0\|^2 = \|x^{(k-1)} - x_*^0\|^2 - 2\alpha_r(x^{(k-1)} - x_*^0)^T A^T SS^T A(x^{(k-1)} - x_*^0)$$
$$+\alpha_r^2 \|A^T SS^T(Ax^{(k-1)} - x_*^0)\|^2.$$

Note that for any $H \succeq 0$, it holds that $\lambda_{\max}(H)H \succeq H^2$. Then we have

$$\|A^T SS^T A(x^{(k-1)} - x_*^0)\|^2 = (x^{(k-1)} - x_*^0)^T(A^T SS^T A)^2(x^{(k-1)} - x_*^0)$$
$$\leqslant \lambda_{\max}(A^T SS^T A)(x^{(k-1)} - x_*^0)^T A^T SS^T A(x^{(k-1)} - x_*^0)$$
$$\leqslant \lambda_{\max}^r(x^{(k-1)} - x_*^0)^T A^T SS^T A(x^{(k-1)} - x_*^0).$$
$$\tag{2.13.4}$$

By $x^{(0)} - x_*^0 = A^\dagger(Ax^{(0)} - b) \in range(A^T)$, $A^T SS^T A(x^{(k-1)} - x_*^0) \in range(A^T)$ and (2.13.3), we can prove that $x^{(0)} - x_*^0 \in range(A^T)$ by induction. Therefore,

$$E_{k-1}[\|x^{(k)} - x_*^0\|^2] \leqslant \|x^{(k-1)} - x_*^0\|^2 - 2\alpha_r(x^{(k-1)} - x_*^0)^T A^T A(x^{(k-1)} - x_*^0)$$
$$+\alpha_r^2\lambda_{\max}^r(x^{(k-1)} - x_*^0)^T A^T A(x^{(k-1)} - x_*^0)$$
$$\leqslant (1 - \alpha_r(2 - \alpha_r\lambda_{\max}^r)\sigma_{min}^2(A)\|x^{(k-1)} - x_*^0\|^2.$$

In the last inequality, we use the facts that $-\alpha_r(2 - \alpha_r\lambda_{\max}^r) < 0$, and for all $u \in range(A^T)$, it holds that $u^T A^T Au \geqslant \sigma_{min}(A)\|u\|^2$. Next, by the law of total expectation, we have

$$E[\|x^{(k)} - x_*^0\|^2] \leqslant \eta_r E[\|x^{(k-1)} - x_*^0\|^2].$$

Unrolling the recurrence yields the result. $\square$

According to Theorem 2.13.2, the best convergence rate ($\eta_r = 1 - \sigma_{\min}^2(A)/\lambda_{\max}^r$) of the BRSI algorithm is achieved when $\alpha_r = 1/\lambda_{\max}^r$. However, the proof of Theorem 2.13.2 uses the worst-case estimates, so the convergence bound may be pessimistic, and in practical applications it may not precisely measure the actual convergence rate of the BRSI algorithm.

Theorem 2.13.3. *Assume that $\epsilon > o$ and $0 < \alpha_r < \frac{2}{(1+\epsilon)\lambda_{max}^r}$. If $Ax = b$ is inconsistent, then for arbitrary $x^{(0)} \in \mathcal{R}^n$, the kth iterate $x^{(k)}$ of the BRSI algorithm satisfies*

$$E[\|x^{(k)} - x_*^0\|^2] \leqslant \eta_\epsilon^k \|x^{(0)} - x_*^0\|^2 + \frac{\alpha_r(1 + 1/\epsilon)\gamma(1 - \eta_\epsilon^k)}{(2 - \alpha_r(1 + \epsilon)\lambda_{\max}^r)\sigma_{\min}^2(A)},$$

where $\eta_\epsilon = 1 - \alpha_r(2 - \alpha_r(1 + \epsilon)\lambda_{\max}^r)\sigma_{\min}^2(A)$ and $\gamma = E[\|A^T SS^T(AA^\dagger b - b)\|^2]$.

Proof. It follows from

$$x^{(k)} - x_*^0 = x^{(k-1)} - x_*^0 - \alpha_r A^T SS^T(Ax^{(k-1)} - b)$$

that

$$\begin{aligned}
\|x^{(k)} - x_*^0\|^2 = {}& \|x^{(k-1)} - x_*^0\|^2 - 2\alpha_r(x^{(k-1)} - x_*^0)^T A^T SS^T A(x^{(k-1)} - b) \\
& + \alpha_r^2 \|A^T SS^T(Ax^{(k-1)} - b)\|^2.
\end{aligned}$$

(2.13.5)

By $Ax_*^0 = AA^\dagger b$, triangle inequality, and Young's inequality, we have

$$\begin{aligned}
\|A^T SS^T A(x^{(k-1)} - b)\|^2 &= \|A^T SS^T(Ax^{(k-1)} - Ax_*^0 + AA^\dagger b - b)\|^2 \\
&\leqslant (\|A^T SS^T A(x^{(k-1)} - x_*^0)\| + \|A^T SS^T(AA^\dagger b - b)\|)^2 \\
&\leqslant (1 + \epsilon)\|A^T SS^T A(x^{(k-1)} - x_*^0)\|^2 + (1 + 1/\epsilon)\|A^T SS^T(AA^\dagger b - b)\|)^2.
\end{aligned}$$

(2.13.6)

By (2.13.4), (2.13.5), (2.13.6), and $A^T A A^\dagger b = A^T b$, we have

$$
\begin{aligned}
& E_{k-1}[\|x^{(k)} - x_*^0\|^2] \\
={} & \|x^{(k-1)} - x_*^0\|^2 - 2\alpha_r (x^{(k-1)} - x_*^0)^T A^T A (x^{(k-1)} - x_*^0) \\
& + \alpha_r^2 E_{k-1}[\|A^T S S^T (A x^{(k-1)} - b)\|^2] \\
\leqslant{} & \|x^{(k-1)} - x_*^0\|^2 + \alpha_r^2 (1 + 1/\epsilon)\gamma \\
& - (2\alpha_r - \alpha_r^2 (1 + \epsilon)\lambda_{\max}^r)(x^{(k-1)} - x_*^0)^T A^T A (x^{(k-1)} - x_*^0) \\
\leqslant{} & (1 - \alpha_r(2 - \alpha_r(1 + \epsilon)\lambda_{\max}^r)\sigma_{\min}^2(A))\|x^{(k-1)} - x_*^0\|^2 + \alpha_r^2(1 + 1/\epsilon)\gamma \\
={} & \eta_\epsilon \|x^{(k-1)} - x_*^0\|^2 + \alpha_r^2(1 + 1/\epsilon)\gamma.
\end{aligned}
$$

Then the expected squared norm of the error can be bounded by

$$
\begin{aligned}
E_{k-1}[\|x^{(k)} - x_*^0\|^2] &\leqslant \eta_\epsilon \|x^{(k-1)} - x_*^0\|^2 + \alpha_r^2(1 + 1/\epsilon)\gamma \\
&\leqslant \eta_\epsilon^k \|x^{(0)} - x_*^0\|^2 + \alpha_r^2(1 + 1/\epsilon)\gamma \sum_{i=0}^{k-1} \eta_\epsilon^i \\
&= \eta_\epsilon^k \|x^{(0)} - x_*^0\|^2 + \frac{\alpha_r^2(1 + 1/\epsilon)\gamma(1 - \eta_\epsilon^k)}{1 - \eta_\epsilon} \\
&= \eta_\epsilon^k \|x^{(0)} - x_*^0\|^2 + \frac{\alpha_r(1 + 1/\epsilon)\gamma(1 - \eta_\epsilon^k)}{(2 - \alpha_r(1 + \epsilon)\lambda_{\max}^r)\sigma_{\min}^2(A)}.
\end{aligned}
$$

This completes the proof. $\qquad\square$

The randomized Kaczmarz algorithm

The RK algorithm[39] is one special case of the BRSI algorithm. Choosing $S = \frac{\|A\|_F}{\|A_{i,:}\|} I_{:,i}$ with probability $\frac{\|A_{i,:}\|^2}{\|A\|_F^2}$ in (2.13.2), we have

$$
E[SS^T] = \|A\|_F^2 E\left[\frac{I_{:,i}(I_{:,i})^T}{\|A_{i,:}\|^2}\right] = \|A\|_F^2 \sum_{i=1}^m \frac{I_{:,i}(I_{:,i})^T}{\|A_{i,:}\|^2} \frac{\|A_{i,:}\|^2}{\|A\|_F^2} = \sum_{i=1}^m I_{:,i}(I_{:,i})^T = I,
$$

and recover the RK iteration

$$
x^{(k)} = x^{(k-1)} - \alpha_r \|A\|_F^2 \frac{A_{i,:} x^{(k-1)} - b_i}{\|A_{i,:}\|^2} (A_{i,:})^T. \tag{2.13.7}
$$

For this case, we have $\lambda_{max}^r = \|A\|_F^2$. Choosing $\alpha_r = 1/\|A\|_F^2$ in Theorem 2.13.2 yields the convergence estimate of[39]:

$$
E[\|x^{(k)} - x_*^0\|^2] \leqslant \left(1 - \frac{\sigma_{\min}^2(A)}{\|A\|_F^2}\right)^k \|x^{(0)} - x_*^0\|^2.
$$

The block row uniform sampling algorithm

The BRUS(l) algorithm is one special case of the BRSI algorithm by using uniform sampling and refers to it as the block row uniform sampling (BRUS) algorithm. The BRUS algorithm is also one special case of the randomized average block Kaczmarz algorithm[66]. Assume $1 \leqslant l \leqslant m$. Let $\mathcal{I}$ denote the set consisting of the uniform sampling of l different numbers of $[m]$. Setting $S = \sqrt{m/l} I_{:,\mathcal{I}}$, we have

$$E[SS^T] = \frac{\frac{m}{l}}{C_m^l} \sum_{\mathcal{I} \subseteq [m], |\mathcal{I}| = l} I_{:,\mathcal{I}}(I_{:,\mathcal{I}})^T = \frac{\frac{m}{l}}{C_m^l} C_{m-1}^{l-1} I = I,$$

and obtain the iteration

$$x^{(k)} = x^{(k-1)} - \alpha_r \frac{m}{l} (A_{\mathcal{I},:})^T (A_{\mathcal{I},:} x^{(k-1)} - b_{\mathcal{I}}).$$

For this case, we have

$$\lambda_{\max}^r = \frac{m}{l} \max_{\mathcal{I} \subseteq [m], |\mathcal{I}| = l} \|A_{\mathcal{I},:}\|^2.$$

By Theorem 2.13.2, the BRUS algorithm can have a faster convergence rate than that of the RK algorithm if there exists $l \in [m]$ satisfying

$$\frac{m}{l} \max_{\mathcal{I} \subseteq [m], |\mathcal{I}| = l} \|A_{\mathcal{I},:}\|^2 \leqslant \|A\|_F^2.$$

If initial guess $x^{(0)} = 0$, the details of the BRUS algorithm with block size l is presented as following Algorithm BRUS(l). Note that the constant m/l is incorporated in the stepsize parameter α_r.

When $S = I$, we refer to the BRSI algorithm as the block column sampling iterative (BCSI) algorithm. Given an arbitrary initial guess $x^{(0)} \in \mathcal{R}^n$, the kth iterate of the BCSI algorithm is

$$x^{(k)} = x^{(k-1)} - \alpha_r TT^T A^T (Ax^{(k-1)} - b), \quad (2.13.8)$$

where the stepsize parameter $\alpha_r > 0$, and the random parameter matrix T is sampled independently in each iteration from a distribution $\mathcal{D}_c$ and

---

Algorithm 2.13.2 BRUS(l) Algorithm

---

procedure $(A, b, x^{(0)} = 0, K, \text{fixed } l, 1 \leqslant l \leqslant m)$

   for $k = 0, 1, \cdots, K$ **do**

      Select randomly a set $\mathcal{I}$ consisting of the uniform sampling of l
      numbers of $[m]$

      Update $x^{(k)} = x^{(k-1)} + \alpha_r(A_{\mathcal{I},:})^T(A_{\mathcal{I},:}x^{(k-1)} - b_{\mathcal{I}})$

   end for

   Output: $x^{(K)}$

end procedure

---

satisfies $E[TT^T] = I$. In the following, we shall present the convergence of $E[\|A(x^{(k)} - A^\dagger b)\|^2]$ for arbitrary linear systems. Throughout, we define

$$\lambda_{\max}^c = \max_{T \sim \mathcal{D}_c} \lambda_{\max}(ATT^TA^T).$$

Theorem 2.13.4. *Assume that* $0 < \alpha_c < 2/\lambda_{max}^c$. *For arbitrary* $x^{(0)} \in \mathcal{R}^n$, *the kth iterate* $x^{(k)}$ *of the BCSI algorithm satisfies*

$$E[\|x^{(k)} - A^\dagger b\|^2] \leqslant \eta_c^k \|A(x^{(0)} - A^\dagger b)\|^2,$$

where $\eta_c = 1 - \alpha_c(2 - \alpha_c\lambda_{max}^c)\sigma_{\min}^2(A)$.

Proof. It follows from (2.13.8) and $A^T A A^\dagger b = A^T b$ that

$$\begin{aligned} A(x^{(k)} - A^\dagger b) &= A(x^{(k-1)} - A^\dagger b - \alpha_c TT^T A^T(Ax^{(k-1)} - b)) \\ &= A(x^{(k-1)} - A^\dagger b) - \alpha_c ATT^T A^T A(x^{(k-1)} - A^\dagger b). \end{aligned}$$

Then we have

$$\begin{aligned} \|A(x^{(k)} - A^\dagger b)\|^2 &= \|A(x^{(k-1)} - A^\dagger b)\|^2 \\ &\quad - 2\alpha_c(x^{(k-1)} - A^\dagger b)^T A^T ATT^T A^T A(x^{(k-1)} - A^\dagger b) \\ &\quad + \alpha_c^2(x^{(k-1)} - A^\dagger b)^T A^T(ATT^T A^T)^2 A(x^{(k-1)} - A^\dagger b). \end{aligned}$$

$$(2.13.9)$$

Note that

$$(x^{(k-1)} - A^\dagger b)^T A^T (ATT^T A^T)^2 A(x^{(k-1)} - A^\dagger b)$$
$$\leqslant \lambda_{\max}(ATT^T A^T)(x^{(k-1)} - A^\dagger b)^T A^T ATT^T A^T A(x^{(k-1)} - A^\dagger b)$$
$$\leqslant \lambda_{\max}^c(x^{(k-1)} - A^\dagger b)^T A^T ATT^T A^T A(x^{(k-1)} - A^\dagger b).$$

$$(2.13.10)$$

By (2.13.9),(2.13.10), and $A(x^{(k-1)} - A^\dagger b) \in range(A)$, we have

$$
\begin{aligned}
E[\|A(x^{(k)} - A^\dagger b)\|^2] &= \|A(x^{(k-1)} - A^\dagger b)\|^2 \\
&\quad - 2\alpha_c(x^{(k-1)} - A^\dagger b)^T A^T AA^T A(x^{(k-1)} - A^\dagger b) \\
&\quad + \alpha_c^2 \lambda_{\max}^c(x^{(k-1)} - A^\dagger b)^T A^T AA^T A(x^{(k-1)} - A^\dagger b) \\
&\leqslant (1 - \alpha_c(2 - \alpha_c \lambda_{\max}^c)\sigma_{\min}^2(A))\|A(x^{(k-1)} - A^\dagger b)\|^2.
\end{aligned}
$$

In the last inequality, we use the facts that $-\alpha_c(2 - \alpha_c \lambda_{\max}^c) < 0$, and for all $u \in range(A)$, it holds that $u^T AA^T u \geqslant \sigma_{\min}^2(A)\|u\|^2$. Next, by the law of total expectation, we have

$$E[\|A(x^{(k)} - A^\dagger b)\|^2] \leqslant \eta_c^k \|A(x^{(0)} - A^\dagger b)\|^2.$$

Unrolling the recurrence yields the result. $\qquad\qquad\qquad\qquad\square$

Remark 2.13.1. *If A has full column rank, then Theorem 2.13.4 implies that $x^{(k)}$ of the BCSI algorithm converges linearly to $A^\dagger b$ in the mean square sense.*

The randomized coordinate descent algorithm

The RCD algorithm [20] is one special case of the BCSI algorithm. Choosing $T = \frac{\|A\|_F}{\|A_{:,j}\|} I_{:,j}$ (2.13.8), we have

$$E[TT^T] = \|A\|_F^2 E\left[\frac{I_{:,j}(I_{:,j})^T}{\|A_{:,j}\|^2}\right] = \|A\|_F^2 \sum_{j=1}^n \frac{I_{:,j}(I_{:,j})^T}{\|A_{:,j}\|^2} \frac{\|A_{:,j}\|^2}{\|A\|_F^2} = \sum_{j=1}^n I_{:,j}(I_{:,j})^T = I,$$

and recover the RCD iteration

$$x^{(k)} = x^{(k-1)} - \alpha_c \|A\|_F^2 \frac{(A_{:,j})^T (Ax^{(k-1)} - b)}{\|A_{:,j}\|^2} I_{:,j}.$$

For this case, we have $\lambda_{\max}^c = \|A\|_F^2$. Choosing $\alpha_c = 1/\|A\|_F^2$ in Theorem 2.13.4 yields the convergence estimate of[20,46]:

$$E[\|x^{(k)} - x^0)_*\|^2] \leqslant \left(1 - \frac{\sigma_{\min}^2(A)}{\|A\|_F^2}\right)^k \|x^{(0)} - x_*^0\|^2.$$

The block column uniform sampling algorithm

[68] proposed one new special case of the BCSI algorithm by using uniform sampling and referred to it as the block column uniform sampling (BCUS) algorithm. Assume $1 \leqslant l \leqslant m$. Let $\mathcal{J}$ denote the set consisting of the uniform sampling of l different numbers of $[n]$. Setting $T = \sqrt{n/l}I_{:,\mathcal{J}}$, we have

$$E[TT^T] = \frac{\frac{n}{l}}{C_n^l} \sum_{\mathcal{J}\subseteq[n],|\mathcal{J}|=l} I_{:,\mathcal{J}}(I_{:,\mathcal{J}})^T = \frac{\frac{n}{l}}{C_n^l}C_{n-1}^{l-1}I = I,$$

and obtain the iteration

$$x^{(k)} = x^{(k-1)} - \alpha_c\frac{n}{l}I_{:,\mathcal{J}}(A_{:,\mathcal{J}})^T(Ax^{(k-1)} - b).$$

For this case, we have

$$\lambda_{\max}^c = \frac{n}{l}\max_{\mathcal{J}\subseteq[n],|\mathcal{J}|=l}\|A_{:,\mathcal{J}}\|^2.$$

By Theorem 2.13.4, the BCUS algorithm can have a faster convergence rate than that of the RCD algorithm if there exists $l \in [n]$ satisfying

$$\frac{n}{l}\max_{\mathcal{J}\subseteq[n],|\mathcal{J}|=l}\|A_{:,\mathcal{J}}\|^2 \leqslant \|A\|_F^2.$$

To avoid entire matrix-vector multiplications, an auxiliary vector $r_k = b - Ax^{(k)}$ is introduced in each iteration of the BCUS algorithm. If initial guess $x^{(0)} = 0$, the details of the BCUS algorithm with block size l is presented as following Algorithm BCUS(l). Note that the constant n/l is incorporated in the stepsize parameter α_c.

The extended block row or column sampling iterative algorithm

---

Algorithm 2.13.3 BCUS(l) Algorithm

---

procedure $(A,\ b,\ x^{(0)} = 0,\ K,\ \text{fixed } l,\ 1 \leqslant l \leqslant n)$

    for $k = 0, 1, \cdots, K$ **do**

        Select randomly a set $\mathcal{J}$ consisting of the uniform sampling of l

            numbers of $[n]$

        Compute $w^{(k)} = \alpha_c (A_{:,\mathcal{J}})^T r_{k-1}$

        Update $x_{\mathcal{J}}^{(k)} = x_{\mathcal{J}}^{(k-1)} + w^{(k)}$ and $r_k = r_{k-1} - A_{:,\mathcal{J}} w^{(k)}$

    end for

    Output: $x^{(K)}$

end procedure

Solving $A^T z = 0$ by the RK algorithm with initial guess $z^{(0)} \in b + range(A)$ produces a sequence $\{z^{(k)}\}$, which converges to $b - AA^\dagger b$ (see, e.g.,[48]). Zouzias and Freris[40] proved that the kth iterate $x^{(k)}$ (which is produced by one RK update for $Ax = b - z^{(k)}$ from $x^{(k-1)}$) of the REK algorithm converges to a solution of $Ax = AA^\dagger b$. Note that any solution of $Ax = AA^\dagger b$ is a solution of $Ax = b$ if it is consistent or a least squares solution of $Ax = b$ if it is inconsistent. Based on the idea of the REK algorithm[40],[68] proposed an extended block row sampling iterative (EBRSI) algorithm.

Given $z^{(0)} \in b + range(A)$ and an arbitrary initial guess $x^{(0)} \in \mathcal{R}^n$, the iterates of the EBRSI algorithm at step k are defined as

$$z^{(k)} = z^{(k-1)} - \alpha_c ATT^T A^T z^{(k-1)}, \tag{2.13.11}$$

$$x^{(k)} = x^{(k-1)} - \alpha_r A^T SS^T (Ax^{(k-1)} - b + z^{(k)}), \tag{2.13.12}$$

where the random parameter matrices S and T are independent, and satisfy

$$E[SS^T] = I, \quad E[TT^T] = I.$$

Theorem 2.13.5. *Assume that* $0 < \alpha_c < 2/\lambda_{\max}^c$ *and* $0 < \alpha_r < 2/\lambda_{\max}^r$. *For arbitrary* $x^{(0)} \in \mathcal{R}^n$, $z^{(0)} \in b + range(A)$, *and* $\epsilon > 0$, *the kth iterate* $x^{(k)}$

of the EBRSI algorithm satisfies

$$E[\|x^{(k)} - x_*^0\|^2] \leqslant (1 + \epsilon)^k \eta_r^k \|x^{(0)} - x_*^0\|^2$$
$$+ (1 + 1/\epsilon)\alpha_r^2 \lambda_{\max}^r \|z^{(0)} - (I - AA^\dagger)b\|^2 \sum_{i=0}^{k-1} \eta_c^{k-i}(1 + \epsilon)^i \eta_r^i,$$

where $\eta_r = 1 - \alpha_r(2 - \alpha_r \lambda_{\max}^r)\sigma_{\min}^2(A)$, $\eta_c = 1 - \alpha_c(2 - \alpha_c \lambda_{\max}^c)\sigma_{\min}^2(A)$.

Remark 2.13.2. *In Theorem 2.13.5, no assumptions about the dimensions or rank of* A *are assumed, and the system* $Ax = b$ *can be consistent or inconsistent. Let* $\eta = max\{\eta_r, \eta_c\}$. *It follows from* $0 < \alpha_c < 2/\lambda_{\max}^c$ *and* $0 < \alpha_r < 2/\lambda_{\max}^r$ *max that* $\eta < 1$. *Assume that* ϵ *satisfies* $(1 + \epsilon)\eta < 1$. *We have*

$$E[\|x^{(k)} - x_*^0\|^2] \leqslant (1+\epsilon)^k \eta^k (\|x^{(0)} - x_*^0\|^2 + (1+\epsilon)\alpha_r^2 \lambda_{\max}^r \|z^{(0)} - (I-AA^\dagger)b\|^2/\epsilon^2),$$

which shows that the EBRSI algorithm converges linearly in the mean square sense to x_*^0 *with the rate* $(1 + \epsilon)\eta$.

Based on the idea of the REGS algorithm[46],[68] proposed an extended block column and row sampling iterative (EBCRSI) algorithm. The convergence analysis of the EBCRSI algorithm is similar to Theorem 2.13.5.

Chapter 3

Sparse Kaczmarz-Type Methods

3.1 Sparse Solution and Sparse Least Square Solution

Consider the sparse solution problem

$$x^* = arg\min\{\|x\|_0 : Ax = b, A \in \mathcal{R}^{m \times n}\} \qquad (3.1.1)$$

and the sparse least square solution problem

$$x^* = arg\min\{\|x\|_0 : argmin\|Ax - b\|\} \qquad (3.1.2)$$

for solving linear problems (1.1.1).

x^* in (3.1.1) and (3.1.2) are hard to be solved. If we consider the minimal l_2 norm solution of following problem

$$x^* = arg\min\{\|x\| : Ax = b, A \in \mathcal{R}^{m \times n}\}, \qquad (3.1.3)$$

or the minimal l_2 norm least squares solution of (1.1.1)

$$x^* = arg\min\{\|x\| : argmin\|Ax - b\|\}, \qquad (3.1.4)$$

then we get $x^* = A^+b$, where A^+ is the Moore-Penrose inverse of A.

Remark 3.1.1. *Though $\|x\|_0$ is called zero norm, it is not suitable for the second of the three properties of norm:*

(1) Positive definite: $\|x\| \geqslant 0$, and $\|x\| = 0 \Leftrightarrow x = 0$;

(2) Homogeneity: $\|kx\| = |k|\|x\|$;

(3) Triangle inequality: $\|x + y\| \leqslant \|x\| + \|y\|$.

The first property is obviously satisfied.

The third property is also suitable. The proof is as follows:

For each component x_i, denote $|x_i|_0 = \begin{cases} 0, & when\ x_i = 0; \\ 1, & when\ x_i \neq 0. \end{cases}$ then

$$\|x + y\|_0 = \sum_{i=0}^{m} |x_i + y_i|_0 \ \leqslant \ \sum_{i=0}^{m} (|x_i|_0 + |y_i|_0)$$
$$= \sum_{i=0}^{m} |x_i|_0 + \sum_{i=0}^{m} |y_i|_0 = \|x\|_0 + \|y\|_0.$$

Remark 3.1.2. *For $A \in \mathcal{R}^{m \times n}$, a pseudoinverse of A is defined as a matrix A^+ satisfying all of the following four criteria, known as the Moore-Penrose conditions:*

(1) $AA^+A = A$;

(2) $A^+AA^+ = A^+$;

(3) $(AA^+)^T = AA^+$

(4) $(A^+A)^T = A^+A$.

Here we involve two major classes of computational techniques for solving sparse approximation problems and focus on the latter.

1. Greedy pursuit. Iteratively refine a sparse solution by successively identifying one or more components that yield the greatest improvement in quality. Related methods: MP, OMP, ROMP, CosaMP, etc.

2. Convex relaxation. Replace the combinatorial problem with a convex optimization problem. Solve the convex program with algorithms that exploit the problem structure.

(1) A fundamental approach to sparse approximation replaces the combinatorial l_0-norm in (3.1.2) with the l_1-norm, yielding convex optimization

problems that admit tractable algorithms:

$$x^* = arg\min\{\|x\|_1 : argmin\|Ax - b\|_2\}. \tag{3.1.5}$$

This is the Basic Pursuit (BP) Method.

(2) The l_0-norm can also be replaced by a pseudo-norm such as l_p-norm $(0 < p < 1)$.

(3) Jack Xin has proposed the using of $l_1 - l_2$ in $min\{\|x\|_1 - \|x\|_2 : argmin\|Ax - b\|_2\}$ (see [70] and the references therein).

In some concrete sense, the l_1-norm is the closest convex function to the l_0-norm, so this "relaxation" is quite natural.

Theorem 3.1.1. *If x is K-Sparse and A satisfies the restricted isometry property (RIP) of order K with constant $\delta = \delta_k < 1$, i.e., $(1 - \delta_k)\|x\|_2^2 \leqslant \|Ax\|^2 \leqslant (1 + \delta_k)\|x\|_2^2$, then*

$$x^* = arg\min\{\|x\|_1 : Ax = b\} = arg\min\{\|x\|_0 : Ax = b\}. \tag{3.1.6}$$

Theorem 3.1.1 has been proved by Candes, Romberg and Tao[71].

Remark 3.1.3. *Bregman Distance with respect to convex functional $J(u)$ between points u and v is defined as*

$$B_J^p(u, v) = J(u) - J(v) - \langle p, u - v \rangle, \tag{3.1.7}$$

where $p \in \partial J(v)$, the subdifferential of J at v.

If $J(u)$ is twice differentiable, then $J(u)$ is convex $\Leftrightarrow \nabla^2 J(u) > 0$,

$$J(u) = J(v) + \nabla J(v)^T(u - v) + \frac{1}{2}(u - v)^T \nabla^2 J(\xi)(u - v),$$

$\xi = v + t(u - v)$, *and t is between 0 and 1. So*

$$J(u) - J(v) - \nabla J(v)^T(u - v) > 0, \ if \ u \neq v.$$

Remark 3.1.4. *Basis Pursuit problem is described as follows (constrained minimization):*

$$x^* = arg\min_x\{\|x\|_1 : \ Ax = b\},$$

and it can be changed into the following unconstrained minimization with
Lagrange Multiple Method

$$x^* = arg\min_x \{\mu\|x\|_1 + \frac{1}{2}\|Ax - b\|_2^2\}.$$

Consider more general problems:

$$x^* = arg\min_x \{J(x) + \frac{1}{2}\|Ax - b\|_2^2\}. \tag{3.1.8}$$

Bregman method solves a sequence of convex problems in the iterative scheme

$$x^{(k+1)} = arg\min_x \left\{B_J^{p^{(k)}}(x, x^{(k)}) + \frac{1}{2}\|Ax - b\|_2^2\right\}, \tag{3.1.9}$$

for $k = 0, 1, \cdots$ starting from $x^{(0)} = 0$ and $p^{(0)} = 0$ (hence, for $k = 0$, one solves the original problem (3.1.8)). For nondifferentiable J such as $\mu\| \cdot \|_1$ and $\mu TV(\cdot)$, $\partial J(x^{(k+1)})$ may contain more than one element, leading to many possible choices of $p^{(k+1)}$. In (3.1.9), each $p^{(k+1)}$ is chosen based on the optimality condition $0 \in \partial J(x^{(k+1)}) - p^{(k)} + A^T(Ax^{(k+1)} - b)$, which yields $p^{(k+1)} = p^{(k)} - A^T(Ax^{(k+1)} - b)$ [72].

In [73], [74] three key results for the sequence $\{x^{(k)}\}$ are proved. First, $\|Ax^{(k)} - b\|$ converges to 0 monotonically, and $x^* = \lim x^{(k)}$ is a solution of $\min\{J(x) : Ax = b\}$; second, for l_1−like functions J, the convergence is finite; and third, assuming that b is a noisy observation of Ax^*, where x^* is the unknown noiseless signal, $x^{(k)}$ monotonically gets closer to x^* in terms of the Bregman distance $B_J^{p^k}(x^*, x^{(k)})$ at least while $\|Ax^{(k)} - b\| \geqslant \|Ax^* - b\|$ (note that $\|Ax^{(k)} - b\|$ decreases monotonically in k). The first two results were proved in the literature of the augmented Lagrangian method.

Thus, we get the following iterative method

$$\begin{cases} x^{(k+1)} = arg\min_x \left\{B_J^{p^{(k)}}(x, x^{(k)}) + \frac{1}{2}\|Ax - b\|_2^2\right\}, \\ p^{(k+1)} = p^{(k)} - A^T(Ax^{(k+1)} - b), \\ x^{(0)} = p^{(0)} = 0. \end{cases} \tag{3.1.10}$$

Interestingly, (3.1.9) can be interpreted as iteratively "adding back the residual". Since $p^{(k)}$ is in the range space of A^T ($p^{(0)} = 0$, $p^{(1)} = p^{(0)} - A^T(Ax^{(1)} - b)$, $p^{(2)} = p^{(1)} - A^T(Ax^{(2)} - b) = A^T \sum_{i=1}^{2}(b - Ax^{(i)})$, $\cdots$.

Assume that $p^{(k)} = A^T v^{(k)}$, where $v^{(k)} = \sum_{i=1}^{k}(b - Ax^{(i)}))$, the linear term $< p^{(k)}, x > = < v^{(k)}, Ax >$, can merge into $\frac{1}{2}\|Ax - b\|^2$,

$$
\begin{aligned}
x^{(k+1)} &= \arg\min_x \left\{ J(x) - J(x^{(k)}) - < p^{(k)}, x - x^{(k)} > + \tfrac{1}{2}\|Ax - b\|_2^2 \right\} \\
&= \arg\min_x \left\{ J(x) - < p^{(k)}, x > + \tfrac{1}{2}\|Ax - b\|_2^2 \right\} \\
&= \arg\min_x \left\{ J(x) - < v^{(k)}, Ax > + \tfrac{1}{2}\|Ax - b\|_2^2 \right\} \\
&= \arg\min_x \left\{ J(x) + \tfrac{1}{2}\|Ax - b - v^{(k)}\|_2^2 \right\}
\end{aligned}
$$

(3.1.11)

Introduce $b^{(k+1)} = b + v^{(k)}$, then $b^{(k)} = b + v^{(k-1)}$, thus $b^{(k+1)} = b^{(k)} + v^{(k)} - v^{(k-1)} = b^{(k)} + (b - Ax^{(k)})$. yielding the equivalent iterative scheme

$$
\begin{cases}
b^{(k+1)} = b^{(k)} + (b - Ax^{(k)}), \\
x^{(k+1)} = \arg\min_x \left\{ J(x) + \dfrac{1}{2}\|Ax - b^{(k+1)}\|_2^2 \right\}, \\
x^{(0)} = b^{(0)} = 0.
\end{cases}
$$

(3.1.12)

At each iteration, the residual $b - Ax^{(k)}$ is added to, rather than subtracted from, $b^{(k)}$.

In particular, when $J(x) = \mu\|x\|_1$ in (3.1.8), the **linearized Bregman method**[74] is obtained by linearizing the last term $\frac{1}{2}\|Ax - b\|^2$ in (3.1.9) into $\frac{1}{2}\|Ax^{(k)} - b\|^2 + < A^T(Ax^{(k)} - b), x - x^{(k)} >$ and adding the l_2-proximity term $\frac{1}{2\alpha}\|x - x^{(k)}\|^2$, yielding the new iterative scheme

$$
x^{(k+1)} = \arg\min_x B_J^{p^{(k)}}(x, x^{(k)}) + < A^T(Ax^{(k)} - b), x > + \frac{1}{2\alpha}\|x - x^{(k)}\|^2.
$$

(3.1.13)

The update formula of p can be derived from the optimality conditions of (3.1.13):

$$
p^{(k+1)} = p^{(k)} - A^T(Ax^{(k)} - b) - \frac{x^{(k+1)} - x^{(k)}}{\alpha} = \cdots
$$

$$= \sum_{i=0}^{k} A^T(b - Ax^{(i)}) - \frac{x^{(k+1)}}{\alpha}. \tag{3.1.14}$$

Then, introduce

$$v^{(k)} = p^{(k+1)} + \frac{x^{(k+1)}}{\alpha} = p^{(k)} + A^T(b - Ax^{(k)}) + \frac{x^{(k)}}{\alpha} = \sum_{i=0}^{k} A^T(b - Ax^{(i)}), \ \forall k. \tag{3.1.15}$$

That is,

$$v^{(k+1)} = v^{(k)} + A^T(b - Ax^{(k+1)}).$$

And from (3.1.13), we have

$$x^{(k+1)} = \arg\min_x \left\{ \mu \|x\|_1 - \langle p^{(k)}, x \rangle + \langle A^T(Ax^{(k)} - b), x \rangle + \frac{1}{2\alpha} \|x - x^{(k)}\|^2 \right\}$$

$$= \arg\min_x \left\{ \mu \|x\|_1 + \frac{1}{2\alpha} \left\| x - \alpha \left(p^{(k)} - A^T(Ax^{(k)} - b) + \frac{x^{(k)}}{\alpha} \right) \right\|^2 \right\}$$

$$= \arg\min_x \left\{ \mu \|x\|_1 + \frac{1}{2\alpha} \|x - \alpha v^{(k)}\|^2 \right\} = \alpha T_\mu(v^{(k)}). \tag{3.1.16}$$

Therefore, we obtain the simplified iterative scheme

$$\begin{cases} x^{(k+1)} = \alpha T_\mu(v^{(k)}), \\ v^{(k+1)} = v^{(k)} + A^T(b - Ax^{(k+1)}), \\ x^{(0)} = v^{(0)} = 0. \end{cases} \tag{3.1.17}$$

where $T_\mu(x) = (t_\mu(x_1), \cdots, t_\mu(x_n))^T$ is the soft threshold operator

$$t_\mu(x_i) = \begin{cases} x_i - \mu, & x_i > \mu \\ 0, & |x_i| \leqslant \mu \\ x_i + \mu, & x_i < -\mu \end{cases} \tag{3.1.18}$$

As a matter of fact, $T_\mu(x) = \arg\min_y \{\mu \|y\|_1 + \frac{1}{2}\|y - x\|_2^2\}$ (see Figure 3.1.1 for its function diagram).

Sometimes (3.1.17) can stagnate, but the stagnation is easily removed by a technique called kicking[75]. It can happen that, over a sequence of

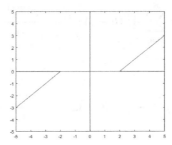

Figure 3.1.1: Soft threshold operator

consecutive iterations, the components v_i satisfying $|v_i| > \mu$ stay constant, while the remaining components v_i, which satisfy $|v_i| \leqslant \mu$, are (slowly) updated. Until one of the latter components finally violates $|v_i| \leqslant \mu$, x remains unchanged. Kicking detected this stagnation by testing $x^{(k)} = x^{(k+1)}$ and broke the stagnation by consolidating all the remaining iterations over which x is unchanged.

It is proved in [76],[77] that the linearized Bregman method converges to the solution of

$$\min \left\{ \mu \|x\|_1 + \frac{1}{2\alpha} \|x\|^2 : Ax = b \right\}. \tag{3.1.19}$$

By scaling the objective function, (3.1.19) can be simplified to

$$\min \left\{ \|x\|_1 + \frac{1}{2\alpha} \|x\|^2 : Ax = b \right\}; \tag{3.1.20}$$

i.e., μ is removed. The convergence was initially established for convex, continuously differentiable convex functions $J(x)$ in [76] (note that both the l_1−norm and total variation must be smoothed to qualify). However, the same paper showed that if the convergence for $J(x) = \|x\|_1$ occurs, then the limit is the solution of (3.1.20). The convergence assumption was later removed in the authors' follow-up paper[77], which was drafted around the same time as the first version of the current paper was written. In addition, it was shown that as $\alpha \to \infty$, the solution of (3.1.20) converges to the solution of the basis pursuitproblem $x^* = argmin\{\|x\|_1 : Ax = b\}$. This paper reduces the requirement to $\alpha > \alpha_\infty$ for a certain finite α_∞.

Remark 3.1.5. *A class of iterative threshold algorithms is formed via different threshold operators.*

$$\begin{cases} v^{(k+1)} = v^{(k)} + \delta A^T(b - Au^{(k)}), \\ u^{(k+1)} = H_{\mu,p}(v^{(k+1)}), \\ u^{(0)} = v^{(0)} = 0, \end{cases} \tag{3.1.21}$$

where $H_{\mu,p}(x) = (h_{\mu,p}(x_1), \cdots, h_{\mu,p}(x_n))$.

Hard threshold operator:

$$H_{\mu,0}(x) = arg\min_x \left\{ \|x\|_1 : arg\min_y \{\mu\|y\|_0 + \frac{1}{2}\|y - x\|_2^2\} \right\},$$

where $H_{\mu,0}(x) = \{h_{\mu,0}(x_1), h_{\mu,0}(x_2), \cdots, h_{\mu,0}(x_n)\}$, and for $i = 1, 2, \cdots, n$

$$h_{\mu,0}(x_i) = \begin{cases} x_i, & |x_i| > \sqrt{2\mu}, \\ 0, & |x_i| \leqslant \sqrt{2\mu}. \end{cases} \tag{3.1.22}$$

Half threshold operator $H_{\mu,\frac{1}{2}} = arg\min_y\{\mu\|y\|_{\frac{1}{2}}^{\frac{1}{2}} + \frac{1}{2}\|y - x\|_2^2\}$, see [78].

$$h_{\mu,\frac{1}{2}}(x_i) = \begin{cases} \frac{2}{3}x_i \left(\cos\left(\frac{2\pi}{3} - \frac{2t_i}{3}\right) + 1\right), & |x_i| > \mu, \\ 0, & |x_i| \leqslant \mu. \end{cases} \tag{3.1.23}$$

where $t_i = \arccos\left(\frac{\sqrt{2}}{2}\left(\frac{\mu}{|x_i|}\right)^{\frac{3}{2}}\right)$.

$2/3-$threshold operator $H_{\mu,\frac{2}{3}} = arg\min_y\{\mu\|y\|_{\frac{2}{3}}^{\frac{2}{3}} + \frac{1}{2}\|y - x\|_2^2\}$, see [78].

$$h_{\mu,\frac{2}{3}} = \begin{cases} sign(x_i)\left(\frac{\phi_\tau(x_i) + \sqrt{\frac{2|x_i|}{\phi_\tau(x_i)} - \phi_\tau(x_i)^2}}{2}\right)^3, & |x_i| > \mu, \\ 0, & |x_i| \leqslant \mu. \end{cases} \tag{3.1.24}$$

where $\phi_\tau = \frac{2^{13/16}}{4\sqrt{3}}\tau^{3/16}\left(\cosh\left(\frac{\theta_\tau(x_i)}{3}\right)\right)^{1/2}$, $\theta_\tau(x_i) = arccosh\left(\frac{3\sqrt{3}x_i^2}{2^{7/4}(2\tau)^{9/8}}\right)$.

3.2 Iterative Hard Thresholding Based on RK

The iterative hard thresholding methods have been recently developed to deal with the sparse regularization problems arising in compressed sensing and other sparse signal processing methods. The methods are attractive due to their simplicity. In 2015, acceleration schemes for iterative hard thresholding methods based on randomized Kaczmarz method was proposed[79]. The speedup is achieved by replacing the gradient step in the iterative hard thresholding methods with a randomized Kaczmarz algorithm.

For solving the l_0 minimizations (3.1.1) or (3.1.2), Blumensath and Davies[80] and[81] introduced the following two minimization schemes:

$$\min_x \mu\|x\|_0 + \|b - Ax\|^2 \tag{3.2.1}$$

and

$$\min_x \|b - Ax\|^2 \ \ subject \ to \ \ \|x\|_0 \leqslant M, \tag{3.2.2}$$

where $\mu > 0$ is a regularization parameter that governs the sparsity of the solution and $M \geqslant 1$ is a given level of sparsity. Furthermore, they suggested using the iterative hard thresholding (IHT) and iterative M-thresholding (IMT) algorithms to solve (3.1.1) and (3.1.2), respectively. The IHT algorithm is defined by the iteration

$$x^{(k+1)} = H_{0,\mu}(x^{(k)} + \alpha A^T(b - Ax^{(k)})). \tag{3.2.3}$$

where $H_{0,\mu}$ is the element-wise hard threshold operator

$$H_{0,\mu}(x_i) = \begin{cases} 0 & if \ |x_i| < \sqrt{\mu}, \\ x_i & if \ |x_i| \geqslant \sqrt{\mu} \end{cases}$$

and $\alpha > 0$ is step size. The IMT algorithm is an iterative procedure of the form

$$x^{(k+1)} = H_M(x^{(k)} + \alpha A^T(b - Ax^{(k)})), \tag{3.2.4}$$

where $H_M : \mathcal{R}^n \to \mathcal{R}^n$ is a M−threshold operator and $H_M(x)$ sets all but M largest magnitude coordinates of x to zero.

To accelerate the convergence of IMT algorithm, Blumensath and Davies[82] suggested using an adaptive size strategy and proposed the normalized iterative $M-$thresholding (NIMT) algorithm as follows:

$$x^{(k+1)} = H_M(x^{(k)} + \alpha_k A^T(b - Ax^{(k)})), \qquad (3.2.5)$$

where the step size α_k are selected by

$$\alpha_k = \frac{\| \left(A^T(b - Ax^{(k)})\right)_{S^k} \|^2}{\|A \left(A^T(b - Ax^{(k)})\right)_{S^k} \|^2}, \quad S^k := supp(x^{(k)}).$$

Here the notation $z_{S^k} \in \mathcal{R}^n$ stands for a vector equal to the vector $z \in \mathcal{R}^n$ in the set S^k and to zero outside of S^k.

The IHT and IMT algorithms are the combination of a gradient descent algorithm to reduce the value of discrepancy $D(x) = \|b - Ax\|^2$ and a hard thresholding operation to reduce the l_0 norm of x. Due to its simplicity, the Kaczmarz method has been a popular and preferable solution tool in many fields, and it is natural to replace the gradient descent step in these two algorithms with a randomized Kaczmarz algorithm to speed up their convergence. The detailed descriptions are presented as follows.

The iteration steps in (3.2.3) and (3.2.5) are iterating along the gradient descent direction. Now consider replacing this part with RK, i.e., using iterative hard threshold operator $H_{0,\mu}$ or H_M after the RK's iteration step (2.1.1).

$$\begin{cases} x_o^{(k+1)} = x^{(k)} + \dfrac{r_i}{\|a_i\|^2}a_i, \\ x^{(k+1)} = H_{0,\mu}(x_o^{(k+1)}). \end{cases} \qquad (3.2.6)$$

$$\begin{cases} x_o^{(k+1)} = x^{(k)} + \dfrac{r_i}{\|a_i\|^2}a_i, \\ x^{(k+1)} = H_M(x_o^{(k+1)}). \end{cases} \qquad (3.2.7)$$

Remark 3.2.1. *What happens if other threshold operators are used instead of hard threshold operator (such as soft threshold operator, half threshold operator $\cdots$)? How about the convergence?*

Remark 3.2.2. *How to change the H_M ?*

Remark 3.2.3. *Can we use the NewRK to replace the RK in this method ?*

Remark 3.2.4. *Can we use the MCK to replace the RK in this method ?*

3.3 Iterative Soft Thresholding Based on RK

In [83], [84], a variant of the Kaczmarz method has been proposed to produce sparse solutions. This sparse Kaczmarz method uses two variables and reads as

$$\begin{cases} v^{(k+1)} & = v^{(k)} + \frac{b_i - \langle a_i, x^{(k)} \rangle}{\|a_i\|^2} a_i, \\ x^{(k+1)} & = T_\mu(v^{(k+1)}), \end{cases} \tag{3.3.1}$$

with $\mu > 0$ and the soft shrinkage function $T_\mu = \max\{|x| - \mu, 0\} \cdot sign(x)$. It has been shown in [83] that for a consistent system $Ax = b$ with an arbitrary matrix A the iterates $x^{(k)}$ converge to the unique solution x^* of the regularized basis pursuit problem,

$$\min_{x \in \mathcal{R}^n} f(\mu) = \mu\|x\|_1 + \frac{1}{2}\|x\|^2, \quad s.t. \ Ax = b, \tag{3.3.2}$$

and for explicit values of $\mu > 0$ guarantee that exact recovery of sparse solutions (also see [85]). The convergence rate has been given in [86].

In [86], authors showed that in the noiseless case the randomized sparse Kaczmarz method in fact converges linearly in expectation without smoothing. And in the noisy case, similarly to the randomized Kaczmarz method, the iterates are expected to reach an error threshold in the order of the noise-level with the same rate as in the noiseless case.

Furthermore, authors extended these results to a more general setting by using the theoretical framework developed in [83]. There the solution x^* of (3.3.2) is considered as a solution of the convex feasibility problem (**CFP**)

$$find \ x \in \mathcal{C} := \bigcap_{i=1}^{m} \mathcal{C}_i \tag{3.3.3}$$

with the hyperplanes $C_i = \{x \in \mathcal{R}^n | \langle a_i, x \rangle = b_i\}$. The Sparse Kaczmarz method is then interpreted as an iterative projection method to solve (3.3.3), where in each iteration, instead of orthogonal projections, Bregman projections with respect to f onto the sets C_i are employed. In this context the iteration with exact Bregman projections reads as

$$v^{(k+1)} = v^{(k)} - t_k \cdot a_i$$

$$x^{(k+1)} = T_\mu(v^{(k+1)}).$$

where t_k is obtained by an exact line-search procedure, and the choice $t_k = \frac{\langle a_i, x^{(k)} \rangle - b_i}{\|a_i\|^2}$ as in (3.3.1) corresponds to an inexact line-search or relaxed Bregman projection. The CFP-framework is quite flexible and allows to include other convex constraints like inequalities $C_i = \{x \in \mathcal{R}^n | \langle a_i, x \rangle \leqslant b_i\}$. Authors showed sublinear convergence rates in expectation for the method of randomized Bregman projections to solve a CFP, where in each iteration an index i is chosen with some probability $p_i > 0$, and a Bregman projection with respect to a general strongly convex function f onto C_i is employed. Moreover, authors derived sufficient conditions which ensure even linear convergence rates. Especially, based on the results in [87], linear rates are obtained for any piecewise linear-quadratic f, and also randomized iterations of the form

$$\begin{cases} V^{(k+1)} &= V^{(k)} + \frac{b_i - \langle A_i, X^{(k)} \rangle}{\|A_i\|_F^2} A_i, \\ X^{(k+1)} &= T_\mu(V^{(k+1)}), \end{cases} \quad (3.3.4)$$

to solve the regularized nuclear norm optimization problem in the area of low rank matrix problems,

$$\min_{X \in \mathcal{R}^{n_1 \times n}} f(X) = \mu\|X\|_* + \frac{1}{2}\|X\|_F^2, \quad s.t. \ \langle A_i, X \rangle = b_i, \ i = 1, \cdots, m, \quad (3.3.5)$$

where $\langle A, X \rangle = trace(A^T X)$ for two matrices $A, X \in \mathcal{R}^{n_1 \times n_2}$, and $T_\mu(X)$ denotes the singular value thresholding operator.

Because of the dual function is needed, we need the following concepts and lemmas. See[88].

Let $f : \mathcal{R}^n \to \mathcal{R}$ be convex. Since f is assumed to be finite everywhere, it is also continuous. By $\partial f(x)$ we denote the subdifferential of f at $x \in \mathcal{R}^n$,

$$\partial f(x) = \{x^* \in \mathcal{R}^n | f(y) \geqslant f(x) + \langle x^*, y - x \rangle \ \ for \ all \ y \in \mathcal{R}^n\},$$

which is nonempty, compact and convex. Furthermore for all $R > 0$ we have

$$\sup_{x \in B_R, \ x^* \in \partial f(x)} \|x^*\| < \infty, \ \ where \ B_R := \{x \in \mathcal{R}^n | \|x\| \leqslant R\}.$$

Definition 3.3.1. *The convex function $f : \mathcal{R}^n \to \mathcal{R}$ is said to be strongly convex, if there is some $\alpha > 0$ such that for all x, $y \in \mathcal{R}^n$ and $x^* \in \partial f(x)$ we have*

$$f(y) \geqslant f(x) + \langle x^*, y - x \rangle + \frac{\alpha}{2}\|y - x\|^2.$$

When the concrete value of α is relevant we indicate this by saying that f is $\alpha-$strongly convex.

Lemma 3.3.1. [88] *If $f : \mathcal{R}^n \to \mathcal{R}$ is $\alpha-$strongly convex then the conjugate function $f^*(y) = \sup_{x \in \mathcal{R}^n} \langle y, x \rangle - f(x)$ is differentiable with a $\frac{1}{\alpha}-$Lipschitz-continuous gradient, i.e.*

$$\|\nabla f^*(y_1) - \nabla f^*(y_2)\| \leqslant \frac{1}{\alpha}\|y_1 - y_2\|, \ \ for \ all \ y_1, y_2 \in \mathcal{R}^n.$$

Example 3.3.1. *The objective function $f(x) = \mu\|x\|_1 + \frac{1}{2}\|x\|^2$ is $1-$strongly convex with $\partial f(x) = \{x + \mu s | s_j = sign(x_j) \ if \ x_j \neq 0, \ and \ s_j \in [-1, 1] \ if \ x_j = 0\}$, $f^*(y) = \frac{1}{2}\|T_\mu(y)\|^2$ and $\nabla f^*(y) = T_\mu(y)$, the soft thresholding operator, cf.[72]. Based on definition (2.1.7), the Bregman distance*

$$D_f^{x^*}(x, y) = \frac{1}{2}\|x - y\|^2 + \mu\left(\|y\|_1 - \langle s, y \rangle\right).$$

Based on the assumption of strong convexity, the following lemma is proved. cf.[83]. It is the key properties of the Bregman distance that are needed for the convergence analysis of the randomized methods.

Lemma 3.3.2. *Let* $f : \mathcal{R}^n \to \mathcal{R}$ *be* $\alpha-$*strongly convex. For all* $x, y \in \mathcal{R}^n$ *and* $x^* \in \partial f(x)$, $y^* \in \partial f(y)$ *we have*

$$\frac{\alpha}{2}\|x - y\|^2 \leqslant D_f^{x^*}(x, y) \leqslant \langle x^* - y^*, x - y \rangle \leqslant \|x^* - y^*\| \cdot \|x - y\|$$

and hence

$$D_f^{x^*}(x, y) = 0 \Leftrightarrow x = y.$$

For sequences x_k *and* $x_k^* \in \partial f(x_k)$, *boundedness of* $D_f^{x_k^*}(x_k, y)$ *implies boundedness of both* x_k *and* x_k^*. *If* f *has a* L-*Lipschitz-continuous gradient then we also have* $D_f(x, y) \leqslant \frac{L}{2}\|x - y\|^2$.

The Bregman projection is defined as follows.

Definition 3.3.2. *Let* $f : \mathcal{R}^n \to \mathcal{R}$ *be strongly convex, and* $\mathcal{C} \subset \mathcal{R}^n$ *be a nonempty closed convex set. The Bregman projection of* x *onto* $\mathcal{C}$ *with respect to* f *and* $p \in \partial f(x)$ *is the unique point* $\Pi_{\mathcal{C}}^p(x)$ *such that*

$$B_f^p\left(x, \Pi_{\mathcal{C}}^p(x)\right) = \min_{y \in \mathcal{C}} B_f^p(x, y) := dist_f^p(x, \mathcal{C})^2.$$

For differentiable f *we simply write* $\Pi_{\mathcal{C}}(x)$ *and* $dist_f(x, \mathcal{C})$.

The notation for the Bregman projection does not capture its dependence on the function f, which, however, will always be clear from the context. Note that for $f(x) = \frac{1}{2}\|x\|^2$ the Bregman projection is just the orthogonal projection onto $\mathcal{C}$ which is denoted by $P_{\mathcal{C}}(x)$. We point out that in this case Bregman distance and the usual distance differ by a factor of 2, but we prefer this slight inconsistency to incorporating the factor into the definition of the Bregman distance. The Bregman projection can also be characterized by a variational inequality.

Lemma 3.3.3. [83] *Let* $f : \mathcal{R}^n \to \mathcal{R}$ *be strongly convex, Then a point* $\hat{x} \in \mathcal{C}$ *is the Bregman projection of* x *onto* $\mathcal{C}$ *with respect to* f *and* $p \in \partial f(x)$ *iff there is some* $\hat{p} \in \partial f(\hat{x})$ *such that one of the following equivalent conditions is fulfilled*

$$\langle \hat{p} - p, y - \hat{x} \rangle \geqslant 0 \quad for \ all \quad y \in \mathcal{C}$$

Algorithm 3.3.1 Randomized Sparse Kaczmarz (RaSK) Algorithm

---

procedure $(A,\ b,\ x^{(0)} = v^{(0)} = 0,\ 0 \neq a_i,\ k = 0)$

while a stopping criterion is not satisfied **do**

Choose an index $i_k \in [m]$ at random with probability $p_{i_k} = \dfrac{\|a_{i_k}\|^2}{\|A\|_F^2}$

Update $v^{(k+1)} = v^{(k)} - \dfrac{\langle a_{i_k}, x^{(k)} \rangle - b_{i_k}}{\|a_{i_k}\|^2} a_{i_k}$

Update $x^{(k+1)} = T_\mu(v^{(k+1)})$

Increment k=k+1

end while

Output: (approximate) solution of $\min\limits_{x \in \mathcal{R}^n} \mu\|x\|_1 + \frac{1}{2}\|x\|^2$ s.t. $Ax = b$

end procedure

$$B_f^{\hat{p}}(\hat{x}, y) \leqslant B_f^p(x, y) - B_f^p(x, \hat{x}) \quad for\ all\ y \in \mathcal{C}.$$

We call any such $\hat{p}$ an admissible subgradient for $\hat{x} = \Pi_{\mathcal{C}}^p(x, \hat{x})$.

Lemma 3.3.4. [83] *Let* $f : \mathcal{R}^n \to \mathcal{R}$ *be* $\alpha-$ *strongly convex,* $A \in \mathcal{R}^{m \times n}$, $b \in \mathcal{R}^m$, $u \in \mathcal{R}^n$ *and* $\beta \in \mathcal{R}$.

(a) The Bregman projection of $x \in \mathcal{R}^n$ *onto the affine subspace* $L(A, b) := \{x \in \mathcal{R}^n \mid Ax = b\} \neq \emptyset$ *is*

$$\hat{x} := \Pi_{L(A,b)}^p(x) = \nabla f^*(p - A^T \hat{w}),$$

where $\hat{w} \in \mathcal{R}^n$ *is a solution of*

$$\min_{w \in \mathcal{R}^m} f^*(p - A^T w) + \langle w, b \rangle.$$

Moreover, $\hat{p} := p - A^T \hat{w}$ *is an admissible subgradient for* x *according to Lemma 3.3.3. If A has full row rank then for all* $y \in L(A, b)$ *we have*

$$B_f^{\hat{p}}(\hat{x}, y) \leqslant B_f^p(x, y) - \frac{\alpha}{2}\|(AA^T)^{-\frac{1}{2}}(Ax - b)\|^2.$$

(b) The Bregman projection of $x \in \mathcal{R}^n$ *onto the hyperplane* $H(u, \beta) := \{x \in \mathcal{R}^n \mid \langle u, x \rangle = \beta\}$ *with* $u \neq 0$ *is*

$$\hat{x} := \Pi_{H(u,\beta)}^p(x) = \nabla f^*(p - \hat{t}u),$$

where $\hat{t} \in \mathcal{R}$ is a solution of

$$\min_{t \in \mathcal{R}} f^*(p - tu) + t\beta.$$

Assume that $b \neq 0$ is in the range $\mathcal{R}(A)$ of A, and hence (3.3.2) has a unique solution $\hat{x} \neq 0$. The Randomized Sparse Kaczmarz Algorithm as follows:

The Exact-step randomized sparse Kaczmarz (ERaSK) method with exact line-search corresponding to a Bregman projection onto the hyperplane $H(a_i, b_i)$ is stated as follows. An exact line-search is indeed computationally feasible, since in this case the derivative of the one-dimensional objective function in line 4 of ERaSK Algorithm is piecewise linear (see [[83], Section 2.5.2]). Computing the "kinks" to determine the linear pieces and then the optimal value can be done with at most $12n$ floating point operations and an $O(n \ln(n))-$ sorting procedure.

Algorithm 3.3.2 Exact-Step Randomized Sparse Kaczmarz (ERaSK) Algorithm

---

procedure $(A,\ b,\ x^{(0)} = v^{(0)} = 0,\ 0 \neq a_i,\ k = 0)$

    while a stopping criterion is not satisfied **do**

        Choose an index $i_k \in [m]$ at random with probability $p_{i_k} = \dfrac{\|a_{i_k}\|^2}{\|A\|_F^2}$

        Calculate $t_k = \arg\min_{t \in \mathcal{R}} f^*(v^{(k)} - ta_{i_k}) + tb_{i_k}$

        Update $v^{(k+1)} = v^{(k)} - t_k a_{i_k}$

        Update $x^{(k+1)} = T_\mu(v^{(k+1)})$

        Increment k=k+1

    end while

    Output: (approximate) solution of $\min_{x \in \mathcal{R}^n} \mu\|x\|_1 + \frac{1}{2}\|x\|^2$ s.t. $Ax = b$

end procedure

Let $supp(\tilde{x}) = \{j \in [n] \,|\, \hat{x}_j \neq 0\}$, denote by $A_{\mathcal{J}}$ the matrix that is formed by the columns of A indexed by $\mathcal{J}$, define

$$\tilde{\sigma}_{min}(A) = \min\{\sigma_{min}(A_{\mathcal{J}}) \,|\, \mathcal{J} \subseteq [n],\ A_{\mathcal{J}} \neq 0\}, \tag{3.3.6}$$

and set $\tilde{\kappa} = \frac{\|A\|_F}{\tilde{\sigma}_{min}(A)}$. In case $b \neq 0$ we also have $\hat{x} \neq 0$ and hence

$$|\hat{x}|_{min} = \min\{|\hat{x}| \,|j \in supp(\hat{x})\,\} > 0. \tag{3.3.7}$$

Now the convergence rate theorem are described as follows.

Theorem 3.3.1. *(noiseless case) The iterates $x^{(k)}$ of the two above algorithms converge in expectation to the unique solution x of the regularized basis pursuit problem (3.3.2) with a linear rate, namely with $\tilde{k} = \frac{\|A\|_F}{\tilde{\sigma}_{min}(A)}$ and contraction factor*

$$q = 1 - \frac{1}{\tilde{k}^2} \cdot \frac{1}{2} \frac{|\hat{x}|_{min}}{|\hat{x}|_{min} + 2\mu} \tag{3.3.8}$$

it holds that

$$E\left[B_f^{v^{(k+1)}}(x^{(k+1)}, \hat{x})\right] \leqslant q \cdot E\left[B_f^{v^{(k)}}(x^{(k)}, \hat{x})\right]$$

and

$$E\left[\|x^{(k)} - \hat{x}\|\right] \leqslant q^{\frac{k}{2}} \sqrt{2\mu\|\hat{x}\|_1 + \|\hat{x}\|^2},$$

where the expectation is taken with respect to the probability distribution $p_i = \frac{\|a_i\|^2}{\|A\|_F^2}$.

Remark 3.3.1. *(a) The contraction factor $q = 1 - \frac{1}{\tilde{k}^2} \cdot \frac{1}{2} \frac{|\hat{x}|_{min}}{|\hat{x}|_{min} + 2\mu}$ depends on $\hat{x}$. This reflects the fact that the dual objective is just restricted strongly convex for $\mu > 0$. The dependence on $\hat{x}$ disappears in the case $\mu = 0$ corresponding to the strongly convex $f(x) = \frac{1}{2}\|x\|^2$.*

(b) If A has full column rank, we have $\tilde{\sigma}_{min}(A) = \sigma_{min}(A)$. Hence for $\mu = 0$ we recover the rate for the standard randomized Kaczmarz method.

Theorem 3.3.2. *(noisy case) Assume that instead of exact data $b \in \mathcal{R}(A)$ only a noisy right hand side $b^\delta \in \mathcal{R}^m$ with $\|b^\delta - b\| \leqslant \delta$ is given. If the iterates $x^{(k)}$ of the both algorithms are computed with b replaced by b^δ, then, with the same contraction factor q from (3.3.8) as in the noiseless case, we* *have*

(a) for the RaSK Algorithm:

$$E\left[\|x^{(k)} - \hat{x}\|\right] \leqslant q^{\frac{k}{2}} \sqrt{2\mu\|\hat{x}\|_1 + \|\hat{x}\|^2} + \sqrt{\frac{2|\hat{x}|_{min} + 4\mu}{|\hat{x}|_{min}}} \cdot \frac{\delta}{\hat{\sigma}_{min}(A)}.$$

(b) for the ERaSK Algorithm with $\|A\|_{1,2} := \sqrt{\sum_{i=1}^{m} \|a_i\|_1^2}$:

$$E\left[\|x^{(k)} - \hat{x}\|\right] \leqslant q^{\frac{k}{2}} \sqrt{2\mu\|\hat{x}\|_1 + \|\hat{x}\|^2}$$
$$+ \sqrt{\frac{2|\hat{x}|_{min} + 4\mu}{|\hat{x}|_{min}}} \cdot \frac{\delta}{\hat{\sigma}_{min}(A)} \sqrt{1 + \frac{4\|A\|_{1,2}}{\delta}}.$$

Remark 3.3.2. *Again we recover the result for the standard randomized Kaczmarz method, because there the step-size* $t_k = \frac{\langle a_{i_k}, x^{(k)}\rangle - b_{i_k}}{\|a_i\|^2}$ *in fact corresponds to an exact line-search. The error threshold for the ERaSK Algorithm is worse than the one for the RaSK Algorithm, which is also observed in numerical experiments. Please note that Theorem 3.3.2 tells us that the RaSK Algorithm and the ERaSK Algorithm are most useful for problems which are almost consistent and are only affected by moderately small noise. This is also the case for the standard randomized Kaczmarz method, which aims at solving the constrained problem* $\min_{x \in \mathcal{R}^n} \|x\|$ *s.t.* $Ax = b$. *If one wishes to compute minimum* $2-norm$ *solutions to unconstrained least squares problems of the form* $\min_{x \in \mathcal{R}^n} \|Ax - b\|$, *then one has to modify the iterations of the standard Kaczmarz method appropriately (see [40]). We do not know yet how to modify the RaSK Algorithm and the ERaSK Algorithm so as to compute minimum* $1-norm$ *solutions to such least squares problems.*

Chapter 4

Nonlinear Kaczmarz-Type Methods and Their Variants

Consider the following systems of algebraic equations

$$f_i(x) = y_i \quad for \quad i = 1, ..., m, \tag{4.0.1}$$

where $f_i : \mathcal{D}(f_i) \subseteq X \to Y$ are possibly nonlinear operators between Hilbert spaces X and Y with domains of definition $\mathcal{D}(f_i)$. Approximate data $y_i^\delta \in Y$ satisfies an estimate of the form $\|y_i^\delta - y_i\| \leqslant \delta_i$ for some noise levels $\delta_i > 0$.

Such problems are often ill-posed and need to be regularized to obtain reasonable approximation solutions. There are at least two basic classes of solution approaches to solve the ill-posed problems (4.0.1), namely (generalized) Tikhonov regularization on the one hand and iterative regularization on the other.

In Tikhonov regularization, one defines approximate solutions as minimizers of the Tikhonov functional $\sum_{i=1}^{m} \|f_i(x) - y_i^\delta\|^2 + \lambda\|x - x_0\|^2$, which is the weighted combination of the residual term $\sum_{i=1}^{m} \|f_i(x) - y_i^\delta\|^2$ that enforces all equations to be approximately satisfied, and the regularization term $\|x - x_0\|^2$ stabilizes the solving process; $\lambda > 0$ is usually referred to as the regularization parameter. In iterative regularization methods, stabilization is achieved via early stopping of iterative schemes. The iteration index in this case plays the role of the regularization parameter, which has to be carefully chosen depending on available information about the noise and the unknowns to be recovered.

For convenience, we abbreviate y_i^δ as y_i, and treat equation (4.0.1) as either consistent (with solution x^*) or inconsistent (without solution).

4.1 Landweber-Kaczmarz-Type Iteration

The most basic iterative regularization method for solving nonlinear ill-posed problems is the Landweber iteration.

4.1.1 Landweber Iteration

The iterative scheme is as follows:

$$x^{(k+1)} = x^{(k)} - \alpha_k \sum_{i=1}^{m} f_i'(x^{(k)})^*(f_i(x^{(k)}) - y_i), \ k = 0, 1, \cdots \qquad (4.1.1)$$

Here $f_i'(x^{(k)})^*$ is the Hilbert space adjoint of the derivative of f_i, α_k is the step size, and $x^{(0)}$ is the initial guess. The Landweber iteration stops the iteration at the smallest index K such that $\sum_{i=1}^{m} \|f_i(x^{(K)}) - y_i\|^2 \leqslant (\tau\delta)^2$ for some constant $\tau > 1$.

In [89], the authors proved the Landweber iteration is a stable method for solving nonlinear ill-posed problems. For perturbed data with noise level δ they proposed a stopping rule that yields the convergence rate $O(\delta^{1/2})$ under appropriate conditions.

Under the general assumptions, the rate of convergence of the Landweber iteration is much slow for linear and nonlinear ill-posed problems.

4.1.2 Landweber-Kaczmarz (LK) Iteration

Each iterative update in (4.1.1) can be numerically quite expensive, since it requires solving forward and adjoint problems for all of the m equations in (4.0.1). In situations where m is large and evaluating the forward and adjoint problems is costly, the classical Landweber-Kaczmarz iteration[90] is

$$x^{(k+1)} = x^{(k)} - f_i'(x^{(k)})^*(f_i(x^{(k)}) - y_i), \ k = 0, 1, \cdots \qquad (4.1.2)$$

where $i = k(mod \ m)+1$, that is, the iteration selects each row of the gradient one by one cyclicly. This method is much faster and is often the method of choice in practice. The acceleration comes from the fact that the update in (4.1.2) only requires the solution of one forward and one adjoint problem instead of solving several of them, but nevertheless often yields a comparable decrease per iteration of the reconstruction error.

Remark 4.1.1. *As is well-known, solving the least square solution of linear system $Ax = b$ is equivalent to solving the following system*

$$A^T A x^* = A^T b \Leftrightarrow x^* = arg \min_x \{\frac{1}{2}x^T A^T A x - x^T A^T b\}.$$

Assume that the step size is α_k, we get gradient descent method for the above linear problem

$$x^{(k+1)} = x^{(k)} + \alpha_k A^T (b - A x^{(k)}), \ k = 0, 1, \cdots.$$

Assume that x^* is the nonlinear least square solution to (4.0.1). Let

$$y = (y_1, \cdots, y_m)^T, \quad f(x) = (f_1(x), \cdots, f_m(x))^T, \quad F(x) = \frac{1}{2}\|f(x) - y\|^2,$$

then

$$x^* = arg \min_x \{\frac{1}{2}\|f(x) - y\|_2^2\} = arg \min_x \{\frac{1}{2}(f(x) - y)^T (f(x) - y)\},$$

and the negative gradient direction

$$-\nabla F(x) = -f'(x)^*(f(x) - y),$$

where

$$f'(x)^*(f(x) - y) = \begin{bmatrix} \frac{\partial f_1}{\partial x_1} & \frac{\partial f_2}{\partial x_1} & \cdots & \frac{\partial f_m}{\partial x_1} \\ \frac{\partial f_1}{\partial x_2} & \frac{\partial f_2}{\partial x_2} & \cdots & \frac{\partial f_m}{\partial x_2} \\ \vdots & \vdots & \ddots & \vdots \\ \frac{\partial f_1}{\partial x_n} & \frac{\partial f_2}{\partial x_n} & \cdots & \frac{\partial f_m}{\partial x_n} \end{bmatrix} \begin{bmatrix} f_1(x) - y_1 \\ f_2(x) - y_2 \\ \vdots \\ f_m(x) - y_m \end{bmatrix}$$

$$= (f_1'(x)^*, f_2'(x)^*, \cdots, f_m'(x)^*) \begin{bmatrix} f_1(x) - y_1 \\ f_2(x) - y_2 \\ \vdots \\ f_m(x) - y_m \end{bmatrix} \tag{4.1.3}$$

$$= \sum_{i=1}^{m} f_i'(x)^*(f_i(x) - y_i).$$

We get a gradient descent method (similar to the Landweber iteration (4.1.1)) for the above nonlinear problem

$$x^{(k+1)} = x^{(k)} - \alpha_k \sum_{i=1}^{m} f_i'(x)^*(f_i(x) - y_i), \quad k = 0, 1, \cdots,$$

where α_k is the step size.

4.1.3　Stochastic Gradient Descent Method

The basic version of stochastic gradient descent (SGD) in $\mathcal{R}^n$ reads as follows: Given the initial guess $x^{(0)} \in \mathcal{R}^n$, update the iterate $x^{(k)}$ by

$$x^{(k+1)} = x^{(k)} - \eta_k f_{i_k}'(x^{(k)})^*(f_{i_k}(x^{(k)}) - y_{i_k}); \quad k = 1, 2, \cdots, \tag{4.1.4}$$

where $f_{i_k}'(x^{(k)}) = f'(x^{(k)})(i_k, :) = \left(\frac{\partial f_{i_k}(x)}{\partial x_1}, \cdots, \frac{\partial f_{i_k}(x)}{\partial x_n} \right)$, that is, the i_kth row of $f'(x^{(k)})$, the index i_k is drawn uniformly from the index set $[m]$ and $\eta_k > 0$ is the corresponding step size[91]. SGD was pioneered by Robbins and Monro in statistical inference[92] (see the monograph[93] for asymptotic convergence results). It has demonstrated encouraging numerical results on diffuse optical tomography[94]. Algorithmically, SGD is a randomized version of the Landweber-Kaczmarz(LK) method (4.1.2).

The theoretical analysis of stochastic iterative methods for inverse problems has just started, and some first theoretical results were obtained in [95, 96] for linear inverse problems. The regularizing property of SGD for linear inverse problems was proved in [96].[91] gave a first thorough analysis of SGD for nonlinear ill-posed inverse problems in the lens of iterative regularization.

To analyze SGD for nonlinear inverse problems, suitable conditions are needed. Since the solution to problem (4.0.1) may be nonunique, the reference solution x^* is taken to be the minimum norm solution (with respect to the initial guess $x^{(0)}$), which is known to be unique under Assumption 4.1.1(ii) below[89].

Assumption 4.1.1. *The following conditions hold:*
(i) The operator $f : \mathcal{R}^n \to \mathcal{R}^m$ is continuous, with a continuous and uniformly bounded Frechet derivative on $\mathcal{R}^n$.
(ii) There exists an $\eta \in (0, 1)$ such that for any x_1, $x_2 \in \mathcal{R}^n$,

$$\|f(x_1) - f(x_2) - f'(x_2)(x_1 - x_2)\| \leqslant \eta\|f(x_1) - f(x_2)\|. \qquad (4.1.5)$$

(iii) There are a family of uniformly bounded operators R_x^i such that for any $x \in \mathcal{R}^n$, $f_i'(x) = R_x^i f_i'(x^*)$ and $R_x = diag(R_x^i) : \mathcal{R}^m \to \mathcal{R}^m$ with

$$\|R_x - I\| \leqslant C_R\|x - x^*\|.$$

(iiii) The source condition holds: There exist some $\nu \in (0, \frac{1}{2})$ and $w \in \mathcal{R}^n$ such that

$$x^* - x^{(0)} = \left(f'(x^*)^* f'(x^*)\right)^\nu w.$$

The conditions in Assumption 4.1.1 are standard for analyzing iterative regularization methods for nonlinear inverse problems[89]. (i) is similar to the L-smoothness commonly used in optimization. (ii)–(iii) have been verified for a class of nonlinear inverse problems[89], e.g., parameter identification for PDEs and nonlinear integral equations. The (4.1.5) is often known as the tangential cone condition (TCC), and it controls the degree of nonlinearity of the operator f. Roughly speaking, it requires that the map f be not far from a linear map. The fractional power $(f'(x^*) * f'(x^*))^\nu$ in (iv) is defined by spectral decomposition (e.g., via the Dunford–Taylor integral). Customarily, it represents a certain smoothness condition on the exact solution x^* (relative to the initial guess $x^{(0)}$). The restriction $\nu < \frac{1}{2}$ is due to technical reasons. It is worth noting that most results require only (i)–(ii), especially the convergence of SGD, whereas (iii)–(iv) are only needed for proving the convergence rate of SGD.

The SGD method often needs suitable assumptions on the step size schedule $\{\eta_k\}_{k=1}^\infty$. The choice is viable since $\max_i \sup_{x \in \mathcal{R}^n} \|f_i'(x)\| < \infty$ by Assumption 4.1.1(i). The choice in Assumption 4.1.2(i) is more general than (ii). The latter choice is often known as a polynomially decaying step size schedule in the literature.

Assumption 4.1.2. *The step sizes $\{\eta_k\}_{k \geqslant 0}$ satisfy one of the following conditions:*

(i) $\eta_k \max_i \sup_{x \in \mathcal{R}^n} \|f_i'(x)\|^2 < 1$ and $\sum_{k=1}^\infty \eta_k = \infty$.

(ii) $\eta_k = \eta_0 k^{-\alpha}$, with $\alpha \in (0,1)$ and $\eta_0 \leqslant \left(\max_i \sup_{x \in \mathcal{R}^n} \|f_i'(x)\|^2 \right)^{-1}$.

In [91], a convergence result gave the regularizing property of SGD for problem (4.0.1) under a priori parameter choice. But the convergence rate is very complex.

Theorem 4.1.1. *(convergence for noisy data y^δ). Let Assumption 4.1.1(i)–(ii) and Assumption 4.1.2(i) be fulfilled. If the stopping index $k(\delta) \in N$ satisfies $\lim_{\delta \to 0^+} k(\delta) = \infty$ and $\lim_{\delta \to 0^+} \delta^2 \sum_{i=1}^{k(\delta)} \eta_i = 0$, then there exists a solution $\tilde{x} \in \mathcal{R}^n$ to problem (4.0.1) such that*

$$\lim_{\delta \to 0^+} E[\|x^{(k(\delta))} - \tilde{x}\|^2] = 0.$$

Further, if $N(F'(x^*)) \subset N(F'(x))$, then

$$\lim_{\delta \to 0^+} E[\|x^{(k(\delta))} - x^*\|^2] = 0.$$

4.1.4   Loping Landweber-Kaczmarz (LLK) Method

In 2007, M. Haltmeier, A. Leitao and O. Scherzer [97] put forward a nov-el iterative regularization technique that consists in considering the classical Landweber-Kaczmarz iteration and incorporating a loping strategy:

$$x^{(k+1)} = x^{(k)} - \omega_k f_i'(x^{(k)})^*(f_i(x^{(k)}) - y_i). \qquad (4.1.6)$$

$$\omega_k = \begin{cases} 1, & \|f_i(x^{(k)}) - y_i\| > \tau\delta_i, \\ 0, & \text{otherwise,} \end{cases} \qquad (4.1.7)$$

where $\tau > 2$ is an appropriate chosen positive constant and $i = k(mod\ m)+1$. The iteration terminates if all ω_k become zero within a cycle, that is if $\|f_i(x^{(k)}) - y_i\| \leqslant \tau\delta_i$ for all $i \in [m]$. This method is referred as the loping Landweber-Kaczmarz method (LLK).

Its worth mentioning that, for noise free data, $\omega_k \equiv 1$ for all k and therefore, in this special situation, the iteration is identical to the classical Landweber-Kaczmarz method.

However, for noisy data, the LLK method is fundamentally different to (4.1.2): the parameter ω_k effects that the iterates defined in (4.1.6) become stationary and all components of the residual vector $\|f_i(x^{(k)}) - y_i\|$ fall below some threshold, making (4.1.6) a convergent regularization method. More-over, especially after a large number of iterations, ω_k will vanish for some k. Therefore, the computational expensive evaluation of $f_i'(x^{(k)})^*$ might be loped, making the Landweber-Kaczmarz method in (4.1.6) a fast alternative to conventional regularization techniques for system of equations.

4.1.5   Embedded Landweber-Kaczmarz (ELK) Method

Also in [97], the authors considered an embedding approach, which consists in rewriting (4.0.1) into an system of equations on the space X^m:

$$f_i(x^i) = y_i, \quad \text{with constraint} \quad \sum_{i=1}^{m} \|x^{i+1} - x^i\|^2 = 0, \qquad (4.1.8)$$

where $x^i \in X^n$, $i = 1, 2, \cdots, m$, and $x^{m+1} := x^1$. Notice that if x is a solution of (4.0.1), then the constant vector $x^i \equiv x$, $i = 1, 2, \cdots, m$ is a solution of system (4.1.8), and vice versa. This system of equations is solved by a block Kaczmarz strategy of the form

$$x^{(k+1/2)} = x^{(k)} - \omega_k f'(x^{(k)})^*(f(x^{(k)}) - y). \tag{4.1.9}$$

$$x^{(k+1)} = x^{(k+1/2)} - \omega_{k+1/2} G(x^{(k+1/2)}), \tag{4.1.10}$$

where $x := (x^i)_i \in X^m$, $y := (y^i)_i \in Y^m$, $f(x) := (f_i(x^i))_i \in Y^m$ and

$$\omega_k = \begin{cases} 1, & \|f(x^{(k)}) - y\| > \tau\delta, \\ 0, & \text{otherwise,} \end{cases} \qquad \omega_{k+1/2} = \begin{cases} 1, & \|G(x^{(k+1/2)})\| > \tau\varepsilon(\delta), \\ 0, & \text{otherwise.} \end{cases}$$
$$\tag{4.1.11}$$

with $\delta := \max \delta^i$. The strictly increasing function $\varepsilon : [0, \infty) \to [0, \infty)$ satisfies $\varepsilon(\delta) \to 0$, as $\delta \to 0$, and guaranties the existence of a finite stopping index. A natural choice is $\varepsilon(\delta) = \delta$. Moreover, up to a positive multiplicative constant, G corresponds to the steepest gradient descent direction of the functional

$$\mathcal{G}(x) := \sum_{i=1}^{m} \|x^{i+1} - x^i\|^2 \tag{4.1.12}$$

on X^m. Notice that (4.1.10) can also be interpreted as a Landweber-Kaczmarz step with respect to the equation

$$\lambda D(x) = 0, \tag{4.1.13}$$

where $D(x) = (x^{i+1} - x^i)_i \in X^m$ and λ is a small positive parameter such that $\|\lambda D\| \leqslant 1$. Since equation (4.0.1) is embedded into a system of equations on a higher dimensional function space we call the resulting regularization technique embedded Landweber-Kaczmarz (ELK) method.[97] analysed ELK method and proved that the ELK method is well posed, convergent and stable.

4.1.6　Loping Steepest-Descent-Kaczmarz Method (LSDK)

In 2008, A. De Cezaro and etc.[98] proposed a loping Steepest-Descent-Kaczmarz method (LSDK) with the steepest-descent method for solving ill-posed problems. This iterative method is defined by

$$x^{(k+1)} = x^{(k)} - \omega_k \alpha_k s_k, \quad s_k = f_i'(x^{(k)})^* (f_i(x^{(k)}) - y_i), \qquad (4.1.14)$$

where

$$\omega_k = \begin{cases} 1, & \|f_i(x^{(k)}) - y_i\| \geqslant \tau \delta_i, \\ 0, & \text{otherwise}, \end{cases} \qquad (4.1.15)$$

and

$$\alpha_k = \begin{cases} \Phi_{rel}(\|s_k\|^2 / \|f_i'(x^{(k)}) s_k\|^2), & \omega_k = 1, \\ \alpha_{\min}, & \omega_k = 0. \end{cases} \qquad (4.1.16)$$

Here, $\alpha_{\min} > 0$, $\tau \in [0, \infty)$ are appropriate chosen numbers, $i = k \ (mod \ m) + 1 \in [m]$, and $x^{(0)} \in \mathcal{R}^n$ is an initial guess, possibly incorporating some a priori knowledge about the exact solution. The function $\Phi_{rel} : (0, \infty) \to (0, \infty)$ defines a sequence of relaxation parameters and is assumed to be continuous, monotonically increasing, bounded by a constant $\alpha_{\max}$, and to satisfy $\Phi_{rel}(s) \leqslant s$.

A. De Cezaro and etc.[98] proved the convergence of the LSDK method and showed this method is an efficient iterative regularization method.

Remark 4.1.2. *A suitable step size can speed up convergence. A large amount of computational resources are wasted in the iteration of the LK, and a jump parameter can solve the problem.*

4.1.7　Regularizing Newton-Kaczmarz Method

In[99], M. Burger and B. Kaltenbacher introduced a class of stabilizing Newton-Kaczmarz methods for nonlinear ill-posed problems and analyzed their convergence and regularization behaviour. Let $f(x) = (f_1(x), \cdots, f_m(x))^T$, $y = (y_1, \cdots, y_m)^T$. In general, for an ill-posed problem, we cannot expect that $f'(x)$ is continuously invertible, and consequently a standard Newton or Gauss-Newton cannot be used. Modified Newton-type methods for solving (4.0.1) have been studied and analyzed in a lot of publications, such as,[100]. Regularization is here achieved by replacing the generally unbounded inverse of $f'(x)$ in the definition of the Newton step by a bounded approximation, defined via a regularizing operator

$$G_\alpha(f'(x)) \approx f'(x)^\dagger,$$

where $f'(x)^\dagger$ denotes the pseudo-inverse of a linear operator $f'(x)$, α is a small regularization paremeter, and G_α satisfies

$$G_\alpha(f'(x))y \to f'(x)^\dagger y, \quad as \ \alpha \to 0 \ \ \forall y \in \mathcal{R}(f'(x)), \tag{4.1.17}$$

and

$$\|G_\alpha(f'(x))\| \leqslant \Phi(\alpha), \tag{4.1.18}$$

Note that, especially in view of operators $f'(x)$ with unbounded inverses, the constant $\Phi(\alpha)$ has to tend to infinity as α goes to zero; we assume w.l.o.g. that $\Phi(\alpha)$ is strictly monotonically decreasing.

Choosing a sequence (α_n) of regularization parameters and applying the bounded operators $G_{\alpha_n}(f'(x^{(k)}))$ in place of $f'(x^{(k)})^{-1}$ in Newton's method results in the iteration

$$x^{(k+1)} = x^{(k)} - G_{\alpha_k}(f'(x^{(k)}))(f(x^{(k)}) - y). \tag{4.1.19}$$

If G_α is defined by Tikhonov regularization

$$G_\alpha(f'(x)) = \left(f'(x)^* f'(x) + \alpha I\right)^{-1} f'(x), \tag{4.1.20}$$

one arrives at the Levenberg-Marquardt method.

A different class of regularized Newton methods emerged from the iteratively regularized Gauss-Newton method (IRGNM),

$$x^{(k+1)} = x^{(0)} - G_{\alpha_k}(f'(x^{(k)})) \left(f(x^{(k)}) - y - f'(x^{(k)})(x^{(k)} - x^{(0)})\right),$$

with (4.1.20), which was first proposed and analyzed by Bakushinskii in [101] and later extended to regularization with general regularization operators G_{α_k} [102]. Here, $\lim_{k\to\infty} \alpha_k = 0$ is an a priori chosen monotonically decreasing sequence of regularization parameters. One observes that in the limiting case $\alpha_k \to 0$ (i.e., $G_{\alpha_k}(f'(x^{(k)})) \to f'(x^{(k)})^\dagger$) also this formulation is equivalent to the usual Newton method.

This method uses the following successive linearization equations to solve equation (4.0.1),

$$f'(x^{(k)})(x - x^{(0)}) = - \left(f(x^{(k)}) - y - f'(x^{(k)})(x^{(k)} - x^{(0)}) \right), \qquad (4.1.21)$$

which is equavilent to

$$f'(x^{(k)})x = f'(x^{(k)})x^{(k)} - \left(f(x^{(k)}) - y \right).$$

In order to make these Newton-type methods applicable to multiple equations (4.0.1), we combine them with a Kaczmarz approach. Starting from an initial guess $x_{0,i}$, we perform a Newton step for the equation $f_i(x) = y_i$, for i from 1 to m, and repeat this procedure in a cyclic manner. One Incorporate the possibility of different regularization methods G^i for each equation in (4.0.1), and use the "overloading" notation

$$x_{0,i} = x_{0,k(mod\ m)+1}, \quad f_i = f_{k(mod\ m)+1}, \quad y_i = y_{k(mod\ m)+1},$$

$$(4.1.22)$$

$$G^i_\alpha = G^{k(mod\ m)+1}_\alpha,$$

write the**Regularized Newton-Kaczmarz method**

$$x^{(k+1)} = x_{0,i} - G^i_{\alpha_k}(f'_i(x^{(k)})) \left(f_i(x^{(k)}) - y_i - f'_i(x^{(k)})(x^{(k)} - x_{0,i}) \right). \quad (4.1.23)$$

We call the above method as **iteratively regularized Gauss-Newton-Kaczmarz (IRGNK) method** if we set

$$G_\alpha(K) = (K^*K + \alpha I)^{-1}K^*.$$

4.1.8 Loping LM-Kaczmarz (LLMK) method

In 2010, J. Baumeister B. Kaltenbacher and A. Leitäo proposed a loping Levenberg-Marquardt-Kaczmarz method (LLMK method) for solving (4.0.1) [103]. This iterative method is defined by

$$x^{(k+1)} = x^{(k)} + w_k h_k, \; k = 0, 1, \cdots, \tag{4.1.24}$$

where

$$h_k = \left(f_i'(x^{(k)})^* f_i'(x^{(k)}) + \alpha I \right)^{-1} f_i'(x^{(k)})^* \left(y_i - f_i(x^{(k)}) \right), \tag{4.1.25}$$

$$w_k = \begin{cases} 1, & if \; \|f_i(x^{(k)}) - y_i\| \geqslant \tau \delta_i, \\ 0, & \text{otherwise.} \end{cases} \tag{4.1.26}$$

Here $\alpha > 0$ is an appropriately chosen number, $i = k(mod \; m) + 1 \in [m]$, $x^{(0)} \in \mathcal{R}^n$ is an initial guess, possibly incorporating some a priori knowledge about the exact solution, and $\tau > 1$ a fixed constant.

The LLMK method consists in incorporating the Kaczmarz strategy into the Levenberg-Marquardt method. The method is a convergent regularization method under some assumptions [103].

4.1.9 Averaged Kaczmarz (AVEK) Method

In 2018, Housen Li and Markus Haltmeier presented the averaged Kaczmarz iteration for solving inverse problems [104]. This method weights the first n steps, which is equivalent to a linear combination of gradients.

$$x^{(k+1)} = \sum_{l=k-n+1}^{k} \omega_{k-l+1} \xi^{(l)} \; for \; k \geqslant n. \tag{4.1.27}$$

$$\xi^{(l)} = x^{(l)} - s_l \alpha_l f_i'(x^{(l)})^* (f_i(x^{(l)}) - y_i). \tag{4.1.28}$$

$$\alpha_l = \begin{cases} 1, & if \; \|f_i(x^{(l)}) - y_i\| \geqslant \tau_i \delta_i, \\ 0, & \text{otherwise,} \end{cases} \tag{4.1.29}$$

where $i = k(mod\ m) + 1$, $x^{(1)}, ..., x^{(n)}$ are user-specified initial values, $\omega_i \geqslant 0$ are fixed weights satisfying $\sum_{i=1}^{n} \omega_i = 1$ and s_i is the step size. As a Kaczmarz iteration, the AVEK iteration only requires evaluating a single gradient $f_i'(x)^*(f_i(x) - y_i)$ per iterative update, which usually is the numerically most expensive part of evaluating. As a Landweber iteration, if every ω_i is positive, each update in AVEK uses information of all equations, which enhances stability. Notice that the Landweber-Kaczmarz iteration is a special case of the general AVEK iteration with $\omega_1 = 1$ and $\omega_i = 0$ for $i \geqslant 2$.

AVEK method can be seen as a hybrid method between the Landweber and the Kaczmarz methods. As a Kaczmarz method, the method only requires evaluation of one direct and one adjoint subproblem per iterative update. On the other hand, similar to the Landweber iteration, this method uses an average over previous auxiliary iterates which increases stability.

In the exact data case, the observed convergence speed (number of cycles versus reconstruction error) of the AVEK turn out to be somewhere between the Kaczmarz method (fastest) and the Landweber method (slowest). A similar behavior has been observed in the noisy data case. In this case, the minimal reconstruction error for the AVEK is slightly smaller than the one of the Kaczmarz method and equal to the Landweber method. The required number of iterations however is less than the one of the Landweber method.

Remark 4.1.3. *This method requires n initial values. Compared with the LK method or the LLK method, it needs to store the iteration result of the first n steps, and the storage capacity is greatly increased. The simplest way to select all weighting factors is to set them as $1/n$.*

Now we state a local tangential cone condition that is used in the following convergent proofs.

Definition 4.1.1. Local Tangential Cone Condition (TCC) [90]. *There exists a ball $B_\rho(x^{(0)}) \subset \mathcal{D}(f_i)$ of radius ρ around $x^{(0)}$, such that $f_i(x)$*

has a locally uniformly bounded Frechet derivative and satisfies the following inequality for any x_1, $x_2 \in B_\rho(x^{(0)})$, $i = 1, 2, \cdots, m$,

$$\|f_i(x_1) - f_i(x_2) - f_i'(x_2)(x_1 - x_2)\| \leqslant \eta_i \|f_i(x_1) - f_i(x_2)\|, \ 0 \leqslant \eta_i < 1. \quad (4.1.30)$$

This condition is often used to ensure that at least an iterative sequence converges locally to a solution x^* of a nonlinear system of equations in $B_{\rho/2}(x^{(0)})$.

We set $x^{(k+1)} = x^{(k)} + \alpha_k s_k$, where $s_k = f_i'(x^{(k)})^*(y_i - f_i(x^{(k)}))$ is the descending direction and α_k is the step length.

If (4.0.1) is consistent, it is easy to get

$$
\begin{aligned}
&\|x^{(k+1)} - x^*\|^2 - \|x^{(k)} - x^*\|^2 \\
&= 2\langle x^{(k)} - x^*, x^{(k+1)} - x^{(k)}\rangle + \|x^{(k+1)} - x^{(k)}\|^2 \\
&= 2\alpha_k \langle x^{(k)} - x^*, f_i'(x^{(k)})^*(y_i - f_i(x^{(k)}))\rangle + \alpha_k^2 \|s_k\|^2 \\
&= 2\alpha_k \langle f_i'(x^{(k)})(x^{(k)} - x^*), y_i - f_i(x^{(k)})\rangle + \alpha_k^2 \|s_k\|^2 \\
&= 2\alpha_k \langle f_i(x^{(k)}) - f_i(x^*) - f_i'(x^{(k)})(x^{(k)} - x^*), f_i(x^{(k)}) - y_i\rangle \\
&\quad - 2\alpha_k \langle f_i(x^{(k)}) - y_i, f_i(x^{(k)}) - y_i\rangle + \alpha_k^2 \|f_i'(x^{(k)})^*(y_i - f_i(x^{(k)}))\|^2 \\
&\leqslant 2\alpha_k(\eta - 1)\|f_i(x^{(k)}) - y_i\|^2 + \alpha_k^2 \|f_i'(x^{(k)})\|^2 \|f_i(x^{(k)}) - y_i\|^2 \\
&= \alpha_k \left(\alpha_k \|f_i'(x^{(k)})\|^2 - 2(1 - \eta)\right) \|f_i(x^{(k)}) - y_i\|^2.
\end{aligned}
\quad (4.1.31)
$$

This ensure the sequences are non-increasing if we assume that $\alpha_k \leqslant \frac{2(1-\eta)}{\|f_i'(x^{(k)})\|^2}$.

If (4.0.1) is inconsistent, we might as well set $f_i(x^*) = y_i + \delta_i$, $i = 1, 2, \cdots, m$, a similar result can be obtained as follows

$$
\begin{aligned}
&\|x^{(k+1)} - x^*\|^2 - \|x^{(k)} - x^*\|^2 \\
&\leqslant \alpha_k \left\{\left(\alpha_k \|f_i'(x^{(k)})\|^2 - 2(1-\eta)\right) \|f_i(x^{(k)}) - y_i\| + 2(1+\eta)\delta_i\right\} \|f_i(x^{(k)}) - y_i\|.
\end{aligned}
\quad (4.1.32)
$$

4.2   Nonlinear Randomized Kaczmarz Method

When y_i is absorbed into function $f_i(x)$, the systems of equations (4.0.1) can be expressed as follows:

$$f_i(x) = 0, \quad i = 1, ..., m, \tag{4.2.1}$$

where $f_i : \mathcal{D}(f_i) \subseteq \mathcal{R}^n \to \mathcal{R}, i = 1, \cdots, m$ are probably nonlinear operators. Suppose that the above system of equations has an isolated solution in a certain region $\mathcal{D}$ (so that the system of equations is consistent). The Newton method is the most commonly used method to solve this problem. But when the system of equations is very large, or Jacobi matrix is singular, the Newton method consumes a lot or can not run.

In consideration of each iterative update of Newton method that can be numerically quite expensive, Kaczmarz-type method is increasingly popular ([90,97–99,103,104]). Using the idea of the randomized Kaczmarz (RK) method, we present a nonlinear orthogonal projection methods: Nonlinear Randomized Kaczmarz (NRK) Method[105].

Note that $f'(x)$ is the Jacobian matrix at x and $\nabla f_i(x)^T$ is ith row, where $f(x) = (f_1(x), \cdots, f_m(x))^T$. The linear approximation can be obtained by the Taylor expansion at $x^{(k)}$ and truncated after the first derivative term :

$$f(x) \approx f(x^{(k)}) + f'(x^{(k)})(x - x^{(k)}). \tag{4.2.2}$$

The approximate solution $x^{(k+1)}$ of $F(x) = 0$ can be approximated by a series of hyperplanes

$$f_i(x^{(k)}) + \langle \nabla f_i(x^{(k)}), x - x^{(k)} \rangle = 0, \quad i = 1, 2, \cdots, m,$$

i. e.,

$$\langle \nabla f_i(x^{(k)}), x \rangle = -f_i(x^{(k)}) + \langle \nabla f_i(x^{(k)}), x^{(k)} \rangle, \quad i = 1, 2, \cdots, m, \tag{4.2.3}$$

where the normal vector of the ith hyperplane is $\nabla f_i(x^{(k)})$. And the directed distance from $x^{(k)}$ to the ith hyperplane is

$$d = \frac{\left(-f_i(x^{(k)}) + \langle \nabla f_i(x^{(k)}), x^{(k)} \rangle\right) - \langle \nabla f_i(x^{(k)}), x^{(k)} \rangle}{\|\nabla f_i(x^{(k)})\|_2} = -\frac{f_i(x^{(k)})}{\|\nabla f_i(x^{(k)})\|_2}.$$

We choose a projection to the ith hyperplane as an iterative improvement of the equations (4.2.3), that is, we define the the the $(k+1)$th iteration as

$$x^{(k+1)} = x^{(k)} - \frac{f_i(x^{(k)})}{\|\nabla f_i(x^{(k)})\|^2} \nabla f_i(x^{(k)}), \qquad (4.2.4)$$

where the choice of i is different from the randomized Kaczmarz (RK) algorithm. To avoid calculating the entire Jacobian matrix for each iteration, the NRK algorithm takes $p_i = \frac{|f_i(x)|^2}{\|f(x)\|^2}$ as the selection probability (preferably selects the row with large error), and only the residual and its norm need to calculate. Therefore, the NRK algorithm reduces a large amount of calculation and storage, which is described in Algorithm 4.2.1.

Algorithm 4.2.1 Nonlinear Randomized Kaczmarz (NRK) Algorithm

---

1: **procedure** ($f, x^{(0)}, K, \varepsilon$)

2:      Initialize $x^{(0)} = x^{(0)}, \ k = 0, \ r = -f(x^{(0)})$

3:     **while** $\|r\| > \varepsilon$ & $k < K$ **do**

4:          Select $i \in [m]$ with probability $\Pr(\text{row}=i) = \frac{|r(i)|^2}{\|r\|^2}$

5:          Compute gradient $g_i = \nabla f_i(x^{(k)})$

6:          Set $x^{(k+1)} = x^{(k)} + \frac{r(i)}{\|g_i\|^2} g_i, \ k = k+1$

7:          Compute residual $r = -f(x^{(k)})$

8:     **end while**

9:      Output $x^{(K)}$

10: **end procedure**

Remark 4.2.1. *Algorithm 4.2.1 has two points completely different from SGD: the first point is that the step size here $\eta_k = 1/\|\nabla f_{i_k}(x^{(k)})\|^2$ is completely determined by the norm of the gradient of the current row; the second point is that the selection probability of the row here is completely determined by the residue.*

In order to prove the convergence of Algorithm 4.2.1, we need to prepare some lemmas and definitions.

Definition 4.2.1. *The matrix $A \in C^{m \times n}$ is called **row bounded below**, if there exists a positive number ϵ such that $\|A(i,:)\| \geqslant \epsilon$, for $1 \leqslant i \leqslant m$.*

Example 4.2.1. *Nonlinear function $f : \mathcal{D} = \{x = (x_1, x_2, x_3)| -1 \leqslant x_i \leqslant 1, i = 1, 2, 3\} \subseteq \mathcal{R}^3 \rightarrow \mathcal{R}^3$:*

$$f(x) = \begin{pmatrix} 3x_1 - \cos(x_2 x_3) - \frac{1}{2} \\ x_1^2 - 81(x_2 + 1)^2 + \sin x_3 + 1.06 \\ e^{-x_1 x_2} + 20x_3 + \frac{10\pi - 3}{3} \end{pmatrix}.$$

Obviously,

$$f'(x) = \begin{bmatrix} 3 & x_3 \sin(x_2 x_3) & x_2 \sin(x_2 x_3) \\ 2x_1 & -162(x_2 + 1) & \cos x_3 \\ -x_2 e^{-x_1 x_2} & -x_1 e^{-x_1 x_2} & 20 \end{bmatrix},$$

and $\|\nabla f_1(x)\|_2 \geqslant 3$, $\|\nabla f_2(x)\|_2 \geqslant \cos(1)$, and $\|\nabla f_3(x)\|_2 \geqslant 20$, therefore, $f'(x)$ is row bounded below.

Lemma 4.2.1. *If the function f satisfies the local tangential cone condition (4.1.30), for $i \in [m]$ and $\forall x_1, x_2 \in \mathcal{R}^n$, we have*

$$\frac{1}{1 + \eta_i} |\nabla f_i(x_1)^T (x_1 - x_2)| \leqslant |f_i(x_1) - f_i(x_2)| \leqslant \frac{1}{1 - \eta_i} |\nabla f_i(x_1)^T (x_1 - x_2)|. \tag{4.2.5}$$

Proof. With the use of the inequality $|a - b| \geqslant |a| - |b|$, we have

$$\eta_i |f_i(x_1) - f_i(x_2)| \geqslant |f_i(x_1) - f_i(x_2) - \nabla f_i(x_1)^T (x_1 - x_2)|$$
$$\geqslant |\nabla f_i(x_1)^T (x_1 - x_2)| - |f_i(x_1) - f_i(x_2)|,$$

therefore,

$$|f_i(x_1) - f_i(x_2)| \geqslant \frac{1}{1 + \eta_i} |\nabla f_i(x_1)^T (x_1 - x_2)|.$$

Similarly,

$$\eta_i |f_i(x_1) - f_i(x_2)| \geqslant |f_i(x_1) - f_i(x_2)| - |\nabla f_i(x_1)^T (x_1 - x_2)|,$$

so

$$|f_i(x_1) - f_i(x_2)| \leqslant \frac{1}{1 - \eta_i} |\nabla f_i(x_1)^T (x_1 - x_2)|.$$

$\square$

Lemma 4.2.2. *If the function f satisfies the local tangential cone condition and $f(x^*) = 0$, from the iteration $x^{(k)} = x^{(k-1)} - \frac{f_i(x^{(k-1)})}{\|\nabla f_i(x^{(k-1)})\|^2} \nabla f_i(x^{(k-1)})$ we have*

$$\|x^{(k)} - x^*\|^2 - \|x^{(k-1)} - x^*\|^2 \leqslant -(1 - 2\eta_i) \frac{|f_i(x^{(k-1)})|^2}{\|\nabla f_i(x^{(k-1)})\|^2}. \qquad (4.2.6)$$

Proof.

$$
\begin{aligned}
&\|x^{(k)} - x^*\|^2 - \|x^{(k-1)} - x^*\|^2 \\
&= \|x^{(k)} - x^{(k-1)}\|^2 + 2\langle x^{(k)} - x^{(k-1)}, x^{(k-1)} - x^* \rangle \\
&= \|\frac{f_i(x^{(k-1)})}{\|\nabla f_i(x^{(k-1)})\|^2} \nabla f_i(x^{(k-1)})\|^2 + 2\langle \frac{-f_i(x^{(k-1)})}{\|\nabla f_i(x^{(k-1)})\|^2} \nabla f_i(x^{(k-1)}), x^{(k-1)} - x^* \rangle \\
&= \frac{|f_i(x^{(k-1)})|^2}{\|\nabla f_i(x^{(k-1)})\|^2} - 2 \frac{f_i(x^{(k-1)})}{\|\nabla f_i(x^{(k-1)})\|^2} \nabla f_i(x^{(k-1)})^T (x^{(k-1)} - x^*) \\
&= \frac{|f_i(x^{(k-1)})|^2}{\|\nabla f_i(x^{(k-1)})\|^2} - 2 \frac{f_i(x^{(k-1)})}{\|\nabla f_i(x^{(k-1)})\|^2} f_i(x^{(k-1)}) \\
&\quad + 2 \frac{f_i(x^{(k-1)})}{\|\nabla f_i(x^{(k-1)})\|^2} (f_i(x^{(k-1)}) - f_i(x^*) - \nabla f_i(x^{(k-1)})^T (x^{(k-1)} - x^*)) \\
&\leqslant \frac{|f_i(x^{(k-1)})|^2}{\|\nabla f_i(x^{(k-1)})\|^2} + 2\eta_i \frac{|f_i(x^{(k-1)})|}{\|\nabla f_i(x^{(k-1)})\|^2} |f_i(x^{(k-1)})| - 2 \frac{|f_i(x^{(k-1)})|^2}{\|\nabla f_i(x^{(k-1)})\|^2} \\
&= -(1 - 2\eta_i) \frac{|f_i(x^{(k-1)})|^2}{\|\nabla f_i(x^{(k-1)})\|^2},
\end{aligned}
$$

$$(4.2.7)$$

where the inequality is due to the local tangential cone condition. $\square$

Lemma 4.2.3. *(Chebyshev's sum inequality) If $a = \{a_1, a_2, \cdots, a_n\}$ and $b = \{b_1, b_2, \cdots, b_n\}$ are two arrays with real components such that $a_1 \geqslant a_2 \geqslant \cdots \geqslant a_n$ and $b_1 \geqslant b_2 \geqslant \cdots \geqslant b_n$, then the following inequality is true*

$$\frac{1}{n} \sum_{j=1}^{n} a_j \sum_{j=1}^{n} b_j \leqslant \sum_{j=1}^{n} a_j b_j. \qquad (4.2.8)$$

Lemma 4.2.4. *If both $a = \{a_1, a_2, \cdots, a_n\}$ and $b = \{b_1, b_2, \cdots, b_n\}$ are two arrays with real components and satisfy $a_j \geqslant 0$, $b_j > 0$, $j \in [n]$, then the following inequality is established*

$$\sum_{j=1}^{n} \frac{a_j}{b_j} \geqslant \frac{\sum_{j=1}^{n} a_j}{\sum_{j=1}^{n} b_j}. \qquad (4.2.9)$$

Proof. (With induction) Obviously, when $n = 1$, the inequality is established. In order to prove that the equation holds when $n = 2$, just prove

$$\frac{a_1}{b_1} + \frac{a_2}{b_2} - \frac{a_1 + a_2}{b_1 + b_2} \geqslant 0.$$

After simplifying the above formula, we get:

$$\frac{a_1 b_2^2 + a_2 b_1^2}{b_1 b_2 (b_1 + b_2)} \geqslant 0,$$

because $a_1 \geqslant 0$, $a_2 \geqslant 0$, $b_1 > 0$, $b_2 > 0$, then the above formula holds.

Suppose the inequality holds when $n = k$, that is

$$\frac{a_1}{b_1} + \frac{a_2}{b_2} + \cdots + \frac{a_k}{b_k} \geqslant \frac{a_1 + a_2 + \cdots + a_k}{b_1 + b_2 + \cdots + b_k},$$

the following part proves that the inequality is established when $n = k + 1$, which can be obtained from the inequality when $n = k$,

$$\frac{a_1}{b_1} + \frac{a_2}{b_2} + \cdots + \frac{a_k}{b_k} + \frac{a_{k+1}}{b_{k+1}} \geqslant \frac{a_1 + a_2 + \cdots + a_k}{b_1 + b_2 + \cdots + b_k} + \frac{a_{k+1}}{b_{k+1}},$$

then from the inequality when $n = 2$ is true, we can get:

$$\frac{a_1}{b_1} + \frac{a_2}{b_2} + \cdots + \frac{a_k}{b_k} + \frac{a_{k+1}}{b_{k+1}} \geqslant \frac{a_1 + a_2 + \cdots + a_k + a_{k+1}}{b_1 + b_2 + \cdots + b_k + b_{k+1}}.$$

$\square$

Theorem 4.2.1. *Consider that the nonlinear system of equations $f(x) = 0$, nonlinear function $f : \mathcal{D} \subseteq \mathcal{R}^n \to \mathcal{R}^m$ on a bounded closed set $\mathcal{D}$, and there exists x^* such that $f(x^*) = 0$. If the derivative of f is continuous in $\mathcal{D}$, and for $\forall x \in \mathcal{D}$, $f'(x)$ is of full column rank and row bounded below; for every*

$i \in [m]$, *nonlinear function* f_i *satisfies the local tangential cone condition,* $\eta = \max_i \eta_i < \frac{1}{2}$, *then the NRK algorithm is convergent and*

$$\boldsymbol{E}\|x^{(k)} - x^*\|^2 \leqslant (1 - \frac{1 - 2\eta}{(1+\eta)^2 m\kappa_F^2(f'(x^{(k-1)}))})\boldsymbol{E}\|x^{(k-1)} - x^*\|^2. \quad (4.2.10)$$

Proof. From Lemma 4.2.2, we can obtain

$$\|x^{(k)} - x^*\|^2 - \|x^{(k-1)} - x^*\|^2 \leqslant -(1 - 2\eta_i)\frac{|f_i(x^{(k-1)})|^2}{\|\nabla f_i(x^{(k-1)})\|^2}. \quad (4.2.11)$$

By taking the conditional expectation on both sides of the above formula, we have:

$$
\begin{aligned}
&E\|x^{(k)} - x^*\|^2 \\
\leqslant\ & E\|x^{(k-1)} - x^*\|^2 - \sum_{i=1}^{m}(1 - 2\eta_i)\frac{|f_i(x^{(k-1)})|^2}{\|f(x^{(k-1)})\|^2}\frac{|f_i(x^{(k-1)})|^2}{\|\nabla f_i(x^{(k-1)})\|^2} \\
\leqslant\ & E\|x^{(k-1)} - x^*\|^2 - \frac{\sum\limits_{i=1}^{m}(1-2\eta_i)|f_i(x^{(k-1)})|^4}{\|f(x^{(k-1)})\|^2\sum\limits_{i=1}^{m}\|\nabla f_i(x^{(k-1)})\|^2} \\
\leqslant\ & E\|x^{(k-1)} - x^*\|^2 - \frac{(1-2\eta)\sum\limits_{i=1}^{m}|f_i(x^{(k-1)})|^2\sum\limits_{i=1}^{m}|f_i(x^{(k-1)})|^2}{m\|f(x^{(k-1)})\|^2\|f'(x^{(k-1)})\|_F^2} \\
=\ & E\|x^{(k-1)} - x^*\|^2 - \frac{(1-2\eta)\sum\limits_{i=1}^{m}|f_i(x^{(k-1)}) - f_i(x^*)|^2}{m\|f'(x^{(k-1)})\|_F^2} \\
\leqslant\ & E\|x^{(k-1)} - x^*\|^2 - \frac{(1-2\eta)\sum\limits_{i=1}^{m}\frac{1}{(1+\eta_i)^2}|\nabla f_i(x^{(k-1)})^T(x^{(k-1)}-x^*)|^2}{m\|f'(x^{(k-1)})\|_F^2} \\
\leqslant\ & E\|x^{(k-1)} - x^*\|^2 - \frac{(1-2\eta)}{(1+\eta)^2}\frac{\sum\limits_{i=1}^{m}|\nabla f_i(x^{(k-1)})^T(x^{(k-1)}-x^*)|^2}{m\|f'(x^{(k-1)})\|_F^2} \\
=\ & E\|x^{(k-1)} - x^*\|^2 - \frac{(1-2\eta)}{(1+\eta)^2}\frac{\|f'(x^{(k-1)})(x^{(k-1)}-x^*)\|^2}{m\|f'(x^{(k-1)})\|_F^2} \\
\leqslant\ & E\|x^{(k-1)} - x^*\|^2 - \frac{(1-2\eta)}{(1+\eta)^2}\frac{\sigma_{\min}^2(f'(x^{(k-1)}))\|x^{(k-1)}-x^*\|^2}{m\|f'(x^{(k-1)})\|_F^2} \\
\leqslant\ & \left(1 - \frac{(1-2\eta)}{(1+\eta)^2 m\kappa_F^2(f'(x^{(k-1)}))}\right) E\|x^{(k-1)} - x^*\|^2.
\end{aligned}
$$

The second, third, and fourth inequalities in the above formula are obtained by Lemma 4.2.4, Lemma 4.2.3, and Lemma 4.2.1, respectively, and in the above formula $\eta = \max_i \eta_i < \frac{1}{2}$, σ_{min} is the smallest non-zero singular value of matrix $f'(x^{(k-1)})$. The sixth inequality uses the singular value inequality of the matrix: $\|Ax\| \geqslant \sigma_{\min}(A)\|x\|$. Since $x^{(k)} \in \mathcal{D}$ (bounded closed set), then $\kappa_F^2(f'(x^{(k-1)})$ is bounded, and the convergence of the NRK algorithm is proved.   □

4.3   NK and NURK

The algorithm ideas and iterative formulas of the NK algorithm and the NURK algorithm are similar to those of the NRK algorithm.The only difference is the choice of rows. The NK algorithm selects the projection rows for each iteration in order, while the NURK algorithm randomly selects rows according to uniformly distributed probability.

Algorithm 4.3.1 Nonlinear Kaczmarz (NK) Algorithm

1: **procedure** ($f, x^{(0)}, K, \varepsilon$)

2:       Initialize $x^{(0)} = x_0, \ k = 0, \ \ r = -f(x^{(0)})$

3:       **while** $\|r\| > \varepsilon$ & $k < K$ **do**

4:             Set $i = k(mod \ m) + 1$

5:             Compute gradient $g_i = \nabla f_i(x^{(k)})$

6:             Set $x^{(k+1)} = x^{(k)} + \frac{r(i)}{\|g_i\|^2} g_i, \ k = k + 1$

7:             Compute residual $r = -f(x^{(k)})$

8:       **end while**

9:       Output $x^{(K)}$

10: **end procedure**

When the number of equations in the nonlinear systems of equations is large, the number of iteration steps will increase accordingly. When the number of iteration steps is large, the iteration process of the NURK algorithm will traverse all the rows. This is roughly the same as the result of the NK algorithm that selects rows in order, because it has the same probability of selecting all rows each time.

Next, we give the expected convergence rate of the NURK method.

Algorithm 4.3.2 Nonlinear Uniformly Randomized Kaczmarz (NURK) Algorithm

---

1: **procedure** (f, x_0, K, ε)

2:     Initialize $x^{(0)} = x_0$, $k = 0$, $r = -f(x^{(0)})$

3:     **while** $\|r\| > \varepsilon$ & $k < K$ **do**

4:         Select an index $i \in [m]$ according to $p_i = unidrnd(m)$

5:         Compute gradient $g_i = \nabla f_i(x^{(k)})$

6:         Set $x^{(k+1)} = x^{(k)} + \frac{r_i}{\|g_i\|^2} g_i$, $k = k + 1$

7:         Compute residual $r = -f(x^{(k)})$

8:     **end while**

9:     Output $x^{(K)}$

10: **end procedure**

Theorem 4.3.1. *Consider that the nonlinear system of equations $f(x) = 0$, nonlinear function $f : \mathcal{D} \subseteq \mathcal{R}^n \to \mathcal{R}^m$ on a bounded closed set closed set $\mathcal{D}$, and there exists x^* such that $f(x^*) = 0$. If the derivative of f is continuous in $\mathcal{D}$, and for $\forall x \in \mathcal{D}$, $f'(x)$ is a row bounded below and full column rank matrix, the nonlinear function f_i satisfies the local tangential cone condition, $\eta = \max_i \eta_i < \frac{1}{2}$, then the NURK algorithm is convergent and*

$$E\|x^{(k)} - x^*\|^2 \leqslant (1 - \frac{1 - 2\eta}{(1 + \eta)^2 m \kappa_F^2(f'(x^{(k-1)}))}) E\|x^{(k-1)} - x^*\|^2. \quad (4.3.1)$$

Proof. Similar to the proof of Theorem 4.2.1, from (4.2.11) we can get

$$\|x^{(k)} - x^*\|^2 - \|x^{(k-1)} - x^*\|^2 \leqslant -(1 - 2\eta_i) \frac{|f(x^{(k-1)})|^2}{\|\nabla f_i(x^{(k-1)})\|^2}. \quad (4.3.2)$$

By taking the full expectation on both sides of the above formula, we have:

$$
\begin{aligned}
E\|x^{(k)} - x^*\|^2
&\leqslant E\|x^{(k-1)} - x^*\|^2 - \sum_{i=1}^{m}(1 - 2\eta_i)\frac{1}{m}\frac{|f(x^{(k-1)})|^2}{\|\nabla f_i(x^{(k-1)})\|^2} \\
&\leqslant E\|x^{(k-1)} - x^*\|^2 - \frac{\sum_{i=1}^{m}(1-2\eta_i)|f_i(x^{(k-1)})|^2}{m\sum_{i=1}^{m}\|\nabla f_i(x^{(k-1)})\|^2} \\
&\leqslant E\|x^{(k-1)} - x^*\|^2 - \frac{\sum_{i=1}^{m}\frac{(1-2\eta_i)}{(1+\eta_i)^2}|\nabla f_i(x^{(k-1)})^T(x^{(k-1)}-x^*)|^2}{m\|f'(x^{(k-1)})\|_F^2} \\
&\leqslant E\|x^{(k-1)} - x^*\|^2 - \frac{(1-2\eta)}{(1+\eta)^2}\frac{\sum_{i=1}^{m}|\nabla f_i(x^{(k-1)})^T(x^{(k-1)}-x^*)|^2}{m\|f'(x^{(k-1)})\|_F^2} \\
&= E\|x^{(k-1)} - x^*\|^2 - \frac{(1-2\eta)}{(1+\eta)^2}\frac{\|f'(x^{(k-1)})x^{(k-1)}-x^*\|^2}{m\|f'(x^{(k-1)})\|_F^2} \\
&\leqslant E\|x^{(k-1)} - x^*\|^2 - \frac{(1-2\eta)}{(1+\eta)^2}\frac{\sigma_{\min}^2(f'(x^{(k-1)}))\|x^{(k-1)}-x^*\|^2}{m\|f'(x^{(k-1)})\|_F^2} \\
&\leqslant \left(1 - \frac{(1-2\eta)}{(1+\eta)^2 m\kappa_F^2(f'(x^{(k-1)}))}\right)E\|x^{(k-1)} - x^*\|^2.
\end{aligned}
$$

The second and the fourth inequalities in the above formula are obtained by Lemma 4.2.4 and Lemma 4.2.1, respectively, and in the above formula $\eta = \max_i \eta_i < \frac{1}{2}$, $\sigma_{\min}(f'(x^{(k-1)}))$ is the the smallest non-zero singular value of $f'(x^{(k-1)})$. The fifth inequality uses the singular value inequality of the matrix: $\|Ax\| \geqslant \sigma_{\min}(A)\|x\|$. Since $x^{(k)} \in \mathcal{D}$ (bounded closed set), so $\kappa_F^2(f'(x^{(k-1)})$ is bounded. The convergence of the NURK algorithm is proved. $\qquad\square$

Now we give the convergence theorem of the NK method.

Assumption 4.3.1. *The following conditions hold:*

(i) The function $f : \mathcal{D} \subseteq \mathcal{R}^n \to \mathcal{R}^m$ is continuous, with a continuous and uniformly bounded Frechét derivative on $\mathcal{D}$.

(ii) **(Local Tangential Cone Condition)** *For every $i \in [m]$ and $\forall x_1$, $x_2 \in \mathcal{R}^n$, there exists $\eta_i \in [0, \eta)(\eta = \max\limits_{1 \leqslant i \leqslant m} \eta_i < \frac{1}{2})$ such that*

$$
|f_i(x_1) - f_i(x_2) - f_i'(x_1)(x_1 - x_2)| \leqslant \eta_i|f_i(x_1) - f_i(x_2)|.
$$

(iii) **(Row Bounded Below)** *There exists a constant $\varepsilon > 0$ such that*

$$
\|\nabla f_i(x)\|^2 \geqslant \varepsilon, \ i \in [m].
$$

(iv) There exists a solution $x^* \in \mathcal{D}$ such that $f(x^*) = 0$ and

$$N(f'(x^*)) \subseteq N(f'(x)), \ \forall \, x \in \mathcal{D}.$$

Remark 4.3.1. *For $\forall x \in \mathcal{D}$, it is easy to see that $f'(x)$ is column full rank implies Assumption 4.3.1 (iv) hold.*

Corollary 4.3.1. *If Assumption 4.3.1 (ii) is fulfilled and there exists $x^* \in \mathcal{D}$ such that $f(x^*) = 0$, for the sequence $\{x^{(k)}\}_{k \geqslant 0}$ in NK method, we have*

$$\|x^{(k+1)} - x^*\|^2 - \|x^{(k)} - x^*\|^2 \leqslant -(1 - 2\eta) \frac{|f_{i_k}(x^{(k)})|^2}{\|\nabla f_{i_k}(x^{(k)})\|^2}, \qquad (4.3.3)$$

$$\sum_{k=0}^{\infty} \frac{|f_{i_k}(x^{(k)})|^2}{\|\nabla f_{i_k}(x^{(k)})\|^2} \leqslant \frac{1}{(1 - 2\eta)} \|x^{(0)} - x^*\|^2 < \infty, \qquad (4.3.4)$$

where $\eta = \max\limits_{1 \leqslant i \leqslant m} \eta_i < \frac{1}{2}$.

Remark 4.3.2. *By Assumption 4.3.1 (i), there exists $M > 0$ such that $\|\nabla f_{i_k}(x)\| \leqslant M, \ \forall i_k \in [m]$. Let $e_k = x^{(k)} - x^*$, and from Corollary 4.3.1, we get*

$$\|x^{(k+1)} - x^*\|^2 \leqslant \|x^{(k)} - x^*\|^2 - \frac{(1 - 2\eta)}{M^2} |f_{i_k}(x^{(k)})|^2.$$

Hence, $\|e_k\|$ is monotonically decreasing. In addition,

$$\sum_{k=0}^{\infty} |f_{i_k}(x^{(k)})|^2 \leqslant M^2 \sum_{k=0}^{\infty} \frac{|f_{i_k}(x^{(k)})|^2}{\|\nabla f_{i_k}(x^{(k)})\|^2} \leqslant \frac{M^2}{(1 - 2\eta)} \|x^{(0)} - x^*\|^2 < \infty,$$

the residual $-f_{i_k}(x^{(k)})$ converges to zero as $k \to \infty$.

Lemma 4.3.1. *If Assumption 4.3.1 (ii) is fulfilled and there exists $x^* \in \mathcal{D}$ such that $f(x^*) = 0$, then the sequence $\{x^{(k)}\}_{k \geqslant 0}$ generated by the NK method is a Cauchy sequence.*

Proof. By Corollary 4.3.1, $\|e_k\|^2$ is monotonically decreasing to some $\zeta \geqslant 0$. Now we prove that the sequence $\{x^{(k)}\}_{k \geqslant 0}$ is a Cauchy sequence.

Let $j, k \in N$ satisfy $k \leqslant j$. Set $k = k_0 \cdot m + (k_1 - 1)$ and $j = j_0 \cdot m + (j_1 - 1)$ with $k_1, j_1 \in [m]$. Let l_0 meet $k_0 \leqslant l_0 \leqslant j_0$ and

$$\sum_{s=1}^{m} |f_s(x^{(l_0 \cdot m + (s-1))})| \leqslant \sum_{s=1}^{m} |f_s(x^{(p_0 \cdot m + (s-1))})|, \text{ for } \forall p_0, \ k_0 \leqslant p_0 \leqslant j_0.$$

(4.3.5)

Let $l_1 := m$ and $l := l_0 \cdot m + (l_1 - 1)$. We have

$$\|e_j - e_k\| \leqslant \|e_j - e_l\| + \|e_l - e_k\|,$$

and

$$\|e_j - e_l\|^2 = 2\langle e_l - e_j, e_l \rangle + \|e_j\|^2 - \|e_l\|^2,$$
$$\|e_l - e_k\|^2 = 2\langle e_l - e_k, e_l \rangle + \|e_k\|^2 - \|e_l\|^2.$$

(4.3.6)

In addition, we can get

$$
\begin{aligned}
|\langle e_l - e_k, e_l \rangle| &= \left| \sum_{t=k}^{l-1} \langle \frac{f_{i_t}(x^{(t)})}{\|\nabla f_{i_t}(x^{(t)})\|^2} \nabla f_{i_t}(x^{(t)}), e_l \rangle \right| \\
&\leqslant \sum_{t=k}^{l-1} \frac{|f_{i_t}(x^{(t)})|}{\|\nabla f_{i_t}(x^{(t)})\|^2} \cdot |\nabla f_{i_t}(x^{(t)})^T (x^{(l)} - x^*)|,
\end{aligned}
$$

(4.3.7)

and

$$
\begin{aligned}
&|\nabla f_{i_t}(x^{(t)})^T (x^{(l)} - x^*)| \\
&= |\nabla f_{i_t}(x^{(i)})^T (x^* - x^{(t)} + x^{(t)} - x^{(l)})| \\
&= |f_{i_t}(x^*) - (f_{i_t}(x^*) - f_{i_t}(x^{(t)}) - \nabla f_{i_t}(x^{(t)})^T (x^* - x^{(t)})) \\
&\quad - f_{i_t}(x^{(l)}) - (f_{i_t}(x^{(t)}) - f_{i_t}(x^{(l)}) - \nabla f_{i_t}(x^{(t)})^T (x^{(t)} - x^{(l)}))| \\
&\leqslant |f_{i_t}(x^*) - f_{i_t}(x^{(l)})| + |f_{i_t}(x^*) - f_{i_t}(x^{(t)}) - \nabla f_{i_t}(x^{(t)})^T (x^* - x^{(t)})| \\
&\quad + |f_{i_t}(x^{(t)}) - f_{i_t}(x^{(l)}) - \nabla f_{i_t}(x^{(t)})^T (x^{(t)} - x^{(l)})| \\
&\leqslant |f_{i_t}(x^*) - f_{i_t}(x^{(l)})| + \eta |f_{i_t}(x^*) - f_{i_t}(x^{(t)})| + \eta |f_{i_t}(x^{(t)}) - f_{i_t}(x^{(l)})| \\
&= |f_{i_t}(x^*) - f_i(x^{(l)})| + \eta |f_{i_t}(x^*) - f_{i_t}(x^{(t)})| \\
&\quad + \eta |f_{i_t}(x^{(t)}) - f_{i_t}(x^*) + f_{i_t}(x^*) - f_{i_t}(x^{(l)})| \\
&\leqslant (1 + \eta)|f_{i_t}(x^*) - f_{i_t}(x^{(l)})| + 2\eta |f_{i_t}(x^*) - f_{i_t}(x^{(t)})| \\
&= (1 + \eta)|f_{i_t}(x^{(l)})| + 2\eta |f_{i_t}(x^{(t)})|.
\end{aligned}
$$

(4.3.8)

Hence,

$$|\langle e_l - e_k, e_l\rangle| \leqslant (1+\eta) \sum_{t=k}^{l-1} \frac{|f_{i_t}(x^{(t)})|}{\|\nabla f_{i_t}(x^{(t)})\|^2} \cdot \|f_{i_t}(x^{(l)})\| + 2\eta \sum_{t=k}^{l-1} \frac{|f_{i_t}(x^{(t)})\|^2}{\|\nabla f_{i_t}(x^{(t)})\|^2}.$$
(4.3.9)

We set $t = i_0 \cdot m + (i_t - 1)$ with $i_t \in [m]$. Using (4.3.5),(4.3.2) and the boundedness and continuity of $f(x)$ and the derivative of $f'(x)$, we find

$$|f_t(x^{(l)})|$$
$$= |f_{i_t}(x^{(l_0 \cdot m + (l_1 - 1))})|$$
$$\leqslant |f_{i_t}(x^{(l_0 \cdot m + (i_t - 1))})| + \sum_{s=i_t}^{l_1-1} |f_{i_t}(x^{(l_0 \cdot m + (s-1))}) - f_{i_t}(x^{(l_0 \cdot m + s)})|$$
$$\leqslant |f_{i_t}(x^{(l_0 \cdot m + (i_t - 1))})|$$
$$+ \frac{1}{1-\eta} \sum_{s=i_t}^{l_1-1} |\nabla f_{i_t}(x^{(l_0 \cdot m + (s-1))})^T(x^{(l_0 \cdot m + (s-1))} - x^{(l_0 \cdot m + s)})|$$
$$\leqslant |f_{i_t}(x^{(l_0 \cdot m + (i_t - 1))})| + \frac{M}{1-\eta} \sum_{s=i_t}^{l_1-1} |(x^{(l_0 \cdot m + (s-1))} - x^{(l_0 \cdot m + s)})|$$
$$\leqslant |f_{i_t}(x^{(l_0 \cdot m + (i_t - 1))})| + \frac{M}{1-\eta} \sum_{s=i_t}^{l_1-1} \|\frac{f_s(x^{(l_0 \cdot m + (s-1))})}{\|\nabla f_s(x^{(l_0 \cdot m + (s-1))})\|^2} \nabla f_s(x^{(l_0 \cdot m + (s-1))})\|$$
$$\leqslant |f_{i_t}(x^{(l_0 \cdot m + (i_t - 1))})| + \frac{M}{\varepsilon(1-\eta)} \sum_{s=i_t}^{l_1-1} |f_s(x^{(l_0 \cdot m + (s-1))})|$$
$$\leqslant \sum_{s=1}^{m} |f_s(x^{(l_0 \cdot m + (s-1))})| + \frac{M}{\varepsilon(1-\eta)} \sum_{s=1}^{m} |f_s(x^{(l_0 \cdot m + (s-1))})|$$
$$= \frac{\varepsilon(1-\eta)+M}{\varepsilon(1-\eta)} \sum_{s=1}^{m} |f_s(x^{(l_0 \cdot m + (s-1))})|$$
(4.3.10)

Using this estimate,

$$\sum_{t=k}^{l-1} |f_{i_t}(x^{(t)})| \cdot |f_{i_t}(x^{(l)})|$$
$$\leqslant \sum_{i_0=\lfloor k/m \rfloor}^{\lfloor (l-1)/m \rfloor} \sum_{i_t=1}^{m} |f_{i_t}(x^{(i_0 \cdot m + (i_t - 1))})| \cdot |f_{i_t}(x^{(l_0 \cdot m + (l_1 - 1))})|$$
$$\leqslant \frac{\varepsilon(1-\eta)+M}{\varepsilon(1-\eta)} \sum_{i_0=\lfloor k/m \rfloor}^{\lfloor (l-1)/m \rfloor} \left(\sum_{i_t=1}^{m} |f_{i_t}(x^{(i_0 \cdot m + (i_t - 1))})| \cdot \sum_{s=1}^{m} |f_s(x^{(l_0 \cdot m + (s-1))})| \right)$$
$$\leqslant \frac{\varepsilon(1-\eta)+M}{\varepsilon(1-\eta)} \sum_{i_0=\lfloor k/m \rfloor}^{\lfloor (l-1)/m \rfloor} \left(\sum_{i_t=1}^{m} |f_{i_t}(x^{(i_0 \cdot m + (i_t - 1))})| \right)^2$$
$$\leqslant m \frac{\varepsilon(1-\eta)+M}{\varepsilon(1-\eta)} \sum_{i_0=\lfloor k/m \rfloor}^{\lfloor (l-1)/m \rfloor} \sum_{i_t=1}^{m} |f_{i_t}(x^{(i_0 \cdot m + (i_t - 1))})|^2$$

$$\leqslant m\frac{\varepsilon(1-\eta)+M}{\varepsilon(1-\eta)} \sum_{t=k-1}^{l-1} |f_{i_t}(x^{(t)})|^2. \tag{4.3.11}$$

Thus,

$$|\langle e_l - e_k, e_l \rangle| \leqslant m(1+\eta)\frac{\varepsilon(1-\eta)+M}{\varepsilon(1-\eta)} \sum_{t=k-1}^{l-1} |f_{i_t}(x^{(t)})|^2 + 2\eta \sum_{t=k-1}^{l-1} |f_{i_t}(x^{(t)})|^2$$
$$= \gamma \sum_{t=k-1}^{l-1} |f_{i_t}(x^{(t)})|^2,$$

$$\tag{4.3.12}$$

where $\gamma = m(1+\eta)\frac{\varepsilon(1-\eta)+M}{\varepsilon(1-\eta)} + 2\eta$.

By the monotone convergence of $\|e_k\|^2$, the last two terms on each of the right hand side of (4.3.6) tend to $\zeta - \zeta = 0$. In addition, the formula (4.3.12) and (4.3.4) imply that the term $\langle e_l - e_k, e_l \rangle$ tends to zero as $k \to \infty$. Likewise, we get $\langle e_l - e_j, e_l \rangle$ also tends to zero as $k \to \infty$. These imply that the right hand side of (4.3.6) tends to zero as $k \to \infty$. Thus the sequences $\{e_k\}_{k\geqslant 0}$ and $\{x^{(k)}\}_{k\geqslant 0}$ are Cauchy sequences.   □

Lemma 4.3.2. *Let Assumptions 4.3.1 (i)-(iii) be fulfilled, then the NK method converges to a solution x^* of problem (4.2.1).*

Proof. By Lemma 4.3.1, $\{x^{(k)}\}_{k\geqslant 0}$ is a Cauchy sequence. Denotes the limit of the sequence by x^*. Since residual $|f_{i_k}(x^{(k)})|^2$ converges to zero as $k \to \infty$ and $i_k = k(mod\ m) + 1$, we find

$$\lim_{k\to\infty} \|f(x^{(k)})\|^2 = \|f(x^*)\| = 0.$$

□

Lemma 4.3.3. *Let Assumptions 4.3.1(i)-(iv) be fulfilled, and $x^{(0)} \in \mathcal{R}(f'(x^*)^T)$, then there holds*

$$x^{(k)} - x^* \in \mathcal{R}(f'(x^*)^T). \tag{4.3.13}$$

In addition, if $f(x) = 0$ has other solutions, then any other solution $\tilde{x}^*$ satisfies

$$x^* - \tilde{x}^* \in N(f'(x^*)), \tag{4.3.14}$$

vice versa.

Proof. By Lemma 4.3.1, $\{x^{(k)}\}_{k \geqslant 0}$ is a Cauchy sequence. Denotes the limit of $\{x^{(k)}\}_{k \geqslant 0}$ by x^*. In addition, from Lemma 4.3.2, the residual $\|f(x^{(k)})\|^2$ converges to zero as $k \to \infty$, so x^* is a solution. For any $i \in [m]$, $\mathcal{R}(\nabla f_i(x^{(k)})^T) \subseteq \mathcal{R}(f'(x^{(k)})^T)$. By the iterative formula $x_{k+1} = x_k - \frac{f_{i_k}(x^{(k)})}{\|\nabla f_{i_k}(x^{(k)})\|^2} \nabla f_{i_k}(x^{(k)})$ and Assumption 4.3.1(iv), we have

$$\mathcal{R}(\nabla f_{i_0}(x^{(0)})^T) \subseteq \mathcal{R}(f'(x^{(0)})^T) = \mathcal{R}(f'(x^*)^T).$$

In addition, $x^{(0)} \in \mathcal{R}(f'(x^*)^T)$, so we get $x^{(1)} \in \mathcal{R}(f'(x^*)^T)$. Likewise, we also get $x^{(2)} \in \mathcal{R}(f'(x^*)^T)$. By analogy, we can get $x^{(k)} \in \mathcal{R}(f'(x^*)^T)$. Since x^* is the limit of $x^{(k)}$, we can obtain $x^* \in \mathcal{R}(f'(x^*)^T)$. Hence, there holds

$$x^{(k)} - x^* \in \mathcal{R}(f'(x^*)^T).$$

By Assumption 4.3.1(ii), we get

$$\frac{1}{1+\eta} \|\nabla f(x_1)(x_1 - x_2)\|_2 \leqslant \|f(x_1) - f(x_2)\|_2 \leqslant \frac{1}{1-\eta} \|\nabla f(x_1)(x_1 - x_2)\|_2$$
(4.3.15)

for $\forall x_1, x_2 \in \mathcal{D}$. If $f : \mathcal{D} \subseteq \mathcal{R}^n \to \mathcal{R}^m$ has other solutions $\tilde{x}^*$. For x^* and $\tilde{x}^*$, from (4.3.14) we get $f'(x^*)(x^* - \tilde{x}^*) = 0$, hence $x^* - \tilde{x}^* \in N(f'(x^*))$. The proof of this lemma is completed. □

In the following, we denote $x^\dagger$ as the unique solution to problem $f(x) = 0$ of minimal distance to $x^{(0)}$.

Theorem 4.3.2. *Let Assumption 1.1 (i)-(iv) be fulfilled. If $x^{(0)} \in \mathcal{R}(f'(x^*)^T)$, then the sequence $\{x^{(k)}\}_{k \geqslant 0}$ generated by the NK method converges to a solution x^* of $f(x) = 0$, that is,*

$$\lim_{k \to \infty} \|x^{(k)} - x^*\|^2 = 0.$$

Further, if $x^\dagger$ denotes the unique solution of minimal distance to $x^{(0)}$, then

$$\lim_{k \to \infty} \|x^{(k)} - x^\dagger\|^2 = 0.$$

Proof. By Lemma 4.3.1, $\{x^{(k)}\}_{k \geqslant 0}$ is a Cauchy sequence. Denotes the limit of the sequence by x^*. From Lemma 4.3.2, the residual $\|f(x^{(k)})\|^2$ converges to zero as $k \to \infty$, so x^* is a solution. Based on Lemma 4.3.3, we notice that $f(x) = 0$ has a unique solution of minimal distance to the initial guess $x^{(0)}$ that satisfies

$$x^\dagger - x^{(0)} \in N(f'(x^\dagger))^\perp,$$

and by Assumption 4.3.1 (iv), we get $N(f'(x^\dagger)) \subseteq N(f'(x^{(k)}))$ for all $k = 0, 1, 2, ...$, so

$$x^{(k)} - x^{(0)} \in N(f'(x^\dagger))^\perp, \quad k = 0, 1, 2, ...$$

then, $x^* - x^{(0)} \in N(f'(x^\dagger))^\perp$. Hence,

$$x^\dagger - x^* = x^\dagger - x^{(0)} + x^{(0)} - x^* \in \mathcal{N}(f'(x^\dagger))^\perp.$$

This and (4.3.14) imply $x^* = x^\dagger$.   □

Remark 4.3.3. *(1). Each iteration of the algorithm 4.3.1 and the algorithm 4.3.2 needs to calculate two vector valued functions $f(x)$ and $\nabla f_i(x)$, and $4n+2m$ computation (4n for $x^{(k+1)}$ and 2m for the norm of r). The amount of calculation is far less than that of one Newton step. It can also be used in the case of overdetermined nonlinear problems ($m \geqslant n$) and the case of singular Jacobian. It is a matrix-free algorithm. (2). The NK method happens to be the Newton method when $m = n = 1$. Therefore, the NK method can be regarded as a natural generalization of the Newton method.*

4.4   Nonlinear Coordinate Descent Method

Consider the system of equations (4.2.1) with some noise. The system has not any solution in a certain region $\mathcal{D}$ (so that the system of equations is inconsistent). So we must consider to solve the following nonlinear least square problem

$$F(x) = \frac{1}{2} \sum_{i-1}^{m} (f_i(x))^2, \qquad (4.4.1)$$

where $f_i : \mathcal{R}^n \to \mathcal{R}$, $i = 1, \cdots, m$ are given functions, and $m \geqslant n$. Now with the use of the idea of CD method for solving linear system of equations, we hope to deduce a nonlinear coordinate descent (NCD) method or Nonlinear Gauss-Seidel method. The NCD method is based on a linear approximation to the components of F (a linear model of F) in the neighbourhood of $x^{(k)}$: We see from the Taylor expansion (4.2.1) that

$$f(x) \approx l(x) \equiv f(x^{(k)}) + f'(x^{(k)})(x - x^{(k)}), \qquad (4.4.2)$$

Inserting this in the definition (4.2.1) of f, we see that

$$F(x) \approx L(x) \equiv \frac{1}{2} l(x)^T l(x) = \frac{1}{2} \| Ax - \left(Ax^{(k)} - b \right) \|^2, \qquad (4.4.3)$$

with $b = f(x^{(k)})$ and $A = f'(x^{(k)}) = \left(\frac{\partial f_i}{\partial x_j} \right) |_{x=x^{(k)}}$. Apply the CD method to $L(x)$, we obtain the following iterative step:

$$x^{(k+1)} = x^{(k)} + \frac{\langle \frac{\partial f}{\partial x_j}, -f(x^{(k)}) \rangle}{\| \frac{\partial f}{\partial x_j} \|^2} e_j, \qquad (4.4.4)$$

where $\frac{\partial f}{\partial x_j} = f'(x^{(k)})(:, j) = \left(\frac{\partial f_1(x)}{\partial x_j}, \cdots, \frac{\partial f_m(x)}{\partial x_j} \right)^T$, that is, the jth column of $f'(x^{(k)})$, and $r^{(k)} = \left(f'(x^{(k)}) x^{(k)} - f(x^{(k)}) \right) - f'(x^{(k)}) x^{(k)} = -f(x^{(k)})$. Thus, we obtain the following algorithm.

Remark 4.4.1. *(1). Each iteration of the algorithm 4.4.1 needs to calculate two vector valued functions $f(x)$ and $\frac{\partial f}{\partial x_j}$, and 6m computation (4m for $x^{(k+1)}$ and 2m for the norm of r). The total computation of one-step iteration is much less than that of Newton's method. It can also be used in the case of overdetermined nonlinear problems with noise and noise-free $(m \geqslant n)$, and the case of singular Jacobian. It is a matrix-free algorithm. (2). The NCD method happens to be the Newton method when $m = n = 1$. Therefore, the NCD method can be regarded as a natural generalization of the Newton method.*

Algorithm 4.4.1 Nonlinear Coordinate Descent (NCD) Algorithm

---

procedure $(f, x^{(0)}, K, \varepsilon)$

  Initialize $x^{(0)} = x^{(0)}$, $k = 0$, $r = -f(x^{(0)})$

  while $\|r\| > \varepsilon$ & $k < K$ **do**

    Set $j = k(mod\ n) + 1$

    Compute $g_j = \frac{\partial f}{\partial x_j}$ and $\alpha_k = \frac{\langle g_j, r \rangle}{\|g_j\|^2}$

    Set $x_j^{(k+1)} = x_j^{(k)} + \alpha_k$, $k = k + 1$

    Compute residual $r = -f(x^{(k)})$

  end while

  Output $x^{(K)}$

end procedure

4.5 Nonlinear Uniformly Randomized CD Method

Nonlinear uniformly random coordinate descent (NURCD) method or nonlinear uniformly randomized Gauss-Seidel (NURGS) method is a randomized version of the nonlinear coordinate descent method (NCD) for the system of equations (4.2.1) with some noise. That is, the system (4.2.1) may not have a solution in a certain region $\mathcal{D}$ (so that the system of equations is inconsistent). So we must consider to solve the nonlinear least square problem (4.4.1).

Each iteration of the NURCD method requires formula (4.4.4), and $j \in [n]$ is chosen at random with uniform probability.

Its convergence proof is similar to the NURK method.

4.6 Nonlinear Maximal Residual Kaczmarz Method

If only the index with the largest residue is selected in each iteration of (4.2.4), we get a new algorithm: Nonlinear Maximal Residual Kaczmarz (NMRK) Algorithm.

Algorithm 4.5.1 Nonlinear Uniformly Randomized CD (NURCD) Algorithm

---

  procedure $(f, x^{(0)}, K, \varepsilon)$

    Initialize $x^{(0)} = x^{(0)}$, $k = 0$, $r = -f(x^{(0)})$

    while $\|r\| > \varepsilon$ & $k < K$ **do**

      Select $j \in [n]$ with uniform probability

      Compute $g_j = \frac{\partial f}{\partial x_j}$ and $\alpha_k = \frac{\langle g_j, r \rangle}{\|g_j\|^2}$

      Set $x^{(k+1)} = x^{(k)} + \alpha_k e_j$, $k = k + 1$

      Compute residual $r = -f(x^{(k)})$

    end while

    Output $x^{(K)}$

  end procedure

Algorithm 4.6.1 Nonlinear Maximal Residual Kaczmarz (NMRK) Algorithm

---

1: **procedure** $(f, x^{(0)}, K, \varepsilon)$

2:    Initialize $x^{(0)} = x^{(0)}$, $k = 0$, $r = -f(x^{(0)})$

3:    **while** $\|r\| > \varepsilon$ & $k < K$ **do**

4:      Select an index $q \in [m]$ with $[*, q] = max(abs(r))$

5:      Compute $g_q = \nabla f_q(x^{(k)})$

6:      Set $x^{(k+1)} = x^{(k)} + \frac{r(q)}{\|g_q\|^2} g_q$, $k = k + 1$

7:      Compute residual $r = -f(x^{(k)})$

8:    **end while**

9:    Output $x^{(K)}$

10: **end procedure**

Theorem 4.6.1. *Consider that the nonlinear system of equations $f(x) = 0$, where nonlinear function $f : \mathcal{D} \subseteq \mathcal{R}^n \to \mathcal{R}^m$ is on a bounded closed set closed set $\mathcal{D}$. Assume there exists x^* such that $f(x^*) = 0$. If the derivative of f is continuous in $\mathcal{D}$, and for $\forall x \in \mathcal{D}$, $f'(x)$ is a row bounded below and full column rank matrix, the nonlinear function f_i satisfies (4.1.30) and $\eta = \max\limits_i \eta_i < \frac{1}{2}$, then the NMRK algorithm is convergent.*

Proof. Similar to the proof of Theorem 4.2.1, we have

$$
\begin{aligned}
&\|x^{(k)} - x^*\|^2 \\
\leqslant\ &\|x^{(k-1)} - x^*\|^2 - (1 - 2\eta_q)\frac{|f_q(x^{(k-1)})|^2}{\|\nabla f_q(x^{(k-1)})\|^2} \\
=\ &\|x^{(k-1)} - x^*\|^2 - (1 - 2\eta_q)\frac{\max\limits_{1\leqslant i\leqslant m}|f_i(x^{(k-1)}) - f_i(x^*)|^2}{\|\nabla f_q(x^{(k-1)})\|^2} \\
\leqslant\ &\|x^{(k-1)} - x^*\|^2 - \frac{1 - 2\eta_q}{\|\nabla f_q(x^{(k-1)})\|^2}\cdot\max\limits_{1\leqslant i\leqslant m}\left(\frac{|\nabla f_i(x^{(k-1)})^T(x^{(k-1)} - x^*)|^2}{(1+\eta_i)^2}\right) \\
\leqslant\ &\|x^{(k-1)} - x^*\|^2 - \frac{1 - 2\eta}{\|f'(x^{(k-1)})\|^2}\cdot\frac{1}{(1+\eta)^2} \\
&\cdot\max\limits_{1\leqslant i\leqslant m}\left(|\nabla f_i(x^{(k-1)})^T(x^{(k-1)} - x^*)|^2\right) \\
\leqslant\ &\|x^{(k-1)} - x^*\|^2 - (1 - 2\eta)\frac{\|f'(x^{(k-1)})(x^{(k-1)} - x^*)\|^2}{m\|f'(x^{(k-1)})\|^2(1+\eta)^2} \\
\leqslant\ &\left(1 - \frac{1 - 2\eta}{m(1+\eta)^2\kappa_F^2(f'(x^{(k-1)}))}\right)\|x^{(k-1)} - x^*\|^2.
\end{aligned}
$$

$$(4.6.1)$$

Therefore, the conclusion of the theorem holds.

4.7 Nonlinear GRK

Adding a greedy strategy to Algorithm 4.2.1, and we can get the following nonlinear greedy randomized Kaczmarz algorithm, NGRK for short.

With the use the previous definitions and some lemmas, we can prove the following theorem.

Theorem 4.7.1. *Consider the nonlinear systems of equations $f(x) = 0$ on a bounded closed set $\mathcal{D}$, where $F : \mathcal{D} \subseteq \mathcal{R}^n \to \mathcal{R}^m$ and there exists $x^* \in \mathcal{D}$ such that $f(x^*) = 0$. If the Jacobi $f'(x)$ is continuous, row bounded below in DD and full column rank matrix, and for every $i \in [m]$, there exit η and*

η_i such that $0 \leqslant \eta = \max\limits_{1 \leqslant i \leqslant m} \eta_i < \frac{1}{2}$ and the local tangential cone condition is satisfied, then the NGRK method is convergent.

Algorithm 4.7.1 Nonlinear GRK (NGRK) Algorithm

1: **procedure** $(f, x^{(0)}, K, \varepsilon)$

2:      Initialize $x^{(0)} = x^{(0)}, \ k = 0, \ r = -f(x^{(0)})$

3:      **while** $\|r\| > \varepsilon \ \& \ k \leqslant K - 1$ **do**

4:          Determine the index of positive integer $\mu_k = \left\{ i_k \mid |r_{i_k}|^2 \geqslant \frac{\|r\|^2}{m} \right\}$

5:          Compute the i-th entry $\tilde{r}_i$ of the vector $\tilde{r}$

$$\tilde{r}_i = \begin{cases} r_i, & \text{if } i \in \mu_k \\ 0, & \text{otherwise} \end{cases}$$

6:          Select $i_k \in \mu_k$ with probability $\Pr(\text{row} = i_k) = \frac{|\tilde{r}_{i_k}|^2}{\|\tilde{r}\|}$

7:          Compute gradient $g_{i_k} = \nabla f_{i_k}(x^{(k)})$

8:          Set $x^{(k+1)} = x^{(k)} + \frac{\tilde{r}_{i_k}}{\|g_{i_k}\|^2} g_{i_k}, \ k = k + 1$

9:          Compute residual $r = -f(x^{(k)})$

10:      **end while**

11:      Output $x^{(K)}$

12: **end procedure**

Proof. $\{x_k\}$ is well-defined. Similar to Lemma 4.2.2, we can get

$$\|x^{(k+1)} - x^*\|^2 \leqslant \|x_k - x^*\|^2 - (1 - 2\eta_{i_k})\frac{|f_{i_k}(x_k)|^2}{\|\nabla f_{i_k}(x_k)\|^2},$$

so

$$E_k\|x^{(k+1)} - x^*\|^2$$

$$\leqslant \|x^{(k)} - x^*\|^2 - (1 - 2\eta_{i_k})\sum_{i\in\mu_k}\frac{|f_i(x^{(k)})|^2}{\sum_{i\in\mu_k}|f_i(x^{(k)})|^2}\frac{|f_i(x^{(k)})|^2}{\|\nabla f_i(x^{(k)})\|^2}$$

$$\leqslant \|x^{(k)} - x^*\|^2 - \frac{(1-2\eta_{i_k})}{m}\sum_{i\in\mu_k}\frac{|f_i(x^{(k)})|^2}{\|\nabla f_i(x^{(k)})\|^2}$$

$$= \|x^{(k)} - x^*\|^2 - \frac{(1-2\eta_{i_k})}{m}\sum_{i\in\mu_k}\frac{|f_i(x^{(k)})-f_i(x^*)|^2}{\|\nabla f_i(x^{(k)})\|^2}$$

$$\leqslant \|x^{(k)} - x^*\|^2 - \frac{(1-2\eta_{i_k})}{m(1+\eta_{i_k})^2}\frac{\sum\limits_{i=1}^{m}\left|\nabla f_i(x^{(k)})^T(x^{(k)}-x^*)\right|^2}{\sum\limits_{i=1}^{m}\|\nabla f_i(x^{(k)})\|^2}$$

$$= \|x^{(k)} - x^*\|^2 - \frac{(1-2\eta_{i_k})}{m(1+\eta_{i_k})^2}\frac{\|f'(x^{(k)})(x^{(k)}-x^*)\|^2}{\|f'(x^{(k)})\|_F^2}$$

$$\leqslant \|x^{(k)} - x^*\|_2^2 - \frac{4(1-2\eta)}{9m\kappa_F^2(f'(x^{(k)}))}\|x^{(k)} - x^*\|^2 \tag{4.7.1}$$

$$= (1 - \frac{4(1-2\eta)}{9m\kappa_F^2(f'(x^{(k)}))})\|x^{(k)} - x^*\|^2.$$

Taking the conditional expectation of the both sides of this equation, and we get

$$E\|x^{(k+1)} - x^*\|_2^2 \leqslant (1 - \frac{4(1-2\eta)}{9m\kappa_F^2(f'(x^{(k)}))})E\|x^{(k)} - x^*\|_2^2, \quad k = 0, 1, 2, ...$$

The second inequality in (4.7.1) uses the definition of μ_k, the third inequality uses Lemma 4.2.4, and the last inequality uses the singular value inequality of matrix $\|Ax\| \geqslant \sigma_{\min}(A)\|x\|$. In addition, $\mathcal{D}$ is the bounded closed set, due to $x^{(k)} \in \mathcal{D}$, $\kappa_F(f'(x^{(k)}))$ is bounded, so the NGRK is convergence. However, the convergence points are related to the initial iteration values.

Chapter 5

Kaczmarz-Type Methods in Other Problems

5.1 Learning Theory for Kaczmarz Methods

5.1.1 Learning Theory for RK

The randomized Kaczmarz algorithm 2.1.1 was generalized in Chen and Powell[106] to a setting with a sequence of independent random measurement vectors $\{\varphi(t) \in \mathcal{R}^n\}_t$ as

$$x^{(k+1)} = x^{(k)} + \frac{y_k - \langle \varphi_k, x^{(k)} \rangle}{\|\varphi_k\|^2} \varphi_k. \tag{5.1.1}$$

When the measurements have no noise $y_k = \langle \varphi_k, x \rangle$, almost sure convergence was proved and quantitative error bounds were provided in [106].

When the linear system $Ax = b$ is overdetermined $(m > n)$ and has no solution, the Kaczmarz algorithm 1.1.1 can be modified by introducing a relaxation parameter $\eta_k > 0$ in front of $\frac{b_i - \langle a_i, x^{(k)} \rangle}{\|a_i\|^2} a_i$ and the output sequence $\{x_k\}$ converges to the least squares solution $\arg\min_{x \in \mathcal{R}^n} \|Ax - b\|^2$ when $\lim_{k \to \infty} \eta_k = 0$ (see, e.g.[8]).

Setting $\psi_k = \frac{\varphi_k}{\|\varphi_k\|} \in S^{n-1}$ and $\tilde{y}_k = \frac{y_k}{\|\varphi_k\|}$ yields an equivalent form of the scheme (5.1.1) as

$$x^{(k+1)} = x^{(k)} + \{\tilde{y}_k - \langle \psi_k, x^{(k)} \rangle\} \psi_k. \tag{5.1.2}$$

This form is similar to those in the literature of online learning for least squares regression. The above form together with the relaxed Kaczmarz method[40] motivates us to consider the following relaxed randomized Kaczmarz algorithm.

Definition 5.1.1. *With normalized measurement vectors $\{\psi_t \in S^{n-1}\}_t$ and sample values $\{\tilde{y}_t \in \mathcal{R}\}_t$, the relaxed randomized Kaczmarz algorithm is defined by*

$$x_{t+1} = x_t + \eta_t \{\tilde{y}_k - \langle \psi_k, x_t \rangle\} \psi_t, \quad t = 0, 1, \cdots, \tag{5.1.3}$$

where $x^{(0)} \in \mathcal{R}^n$ is an initial vector and $\{\eta_t\}$ is a sequence of relaxation parameters or step sizes.

In 2015, Junhong Lin and Dingxuan Zhou[107] provided a learning theory analysis for the relaxed randomized Kaczmarz algorithm in the assumption of $0 < \eta_t \leqslant 2$ for each $t \in N$ and that the sequence $\{z_t := (\psi_t, \tilde{y}_t)\}_{t \in N}$ is independently drawn according to a Borel probability measure ρ on $\mathcal{Z} := S^{n-1} \times \mathcal{R}$ which satisfies $E[|\tilde{y}|^2] < \infty$.

Main Results

To introduce the learning theory approach to the relaxed randomized Kaczmarz algorithm (5.1.3), we can decompose the probability measure ρ on $\mathcal{Z} = S^{n-1} \times \mathcal{R}$ into its marginal distribution ρ_X on $X := S^{n-1}$ and conditional distributions $\rho(\cdot|\psi)$ at $\psi \in X$. The conditional means define the regression function $f_\rho : X \to \mathcal{R}$ as

$$f_\rho(\psi) = \int_{\mathcal{R}} \tilde{y} d\rho(\tilde{y}|\psi), \quad \psi \in X. \tag{5.1.4}$$

The hypothesis space for the Kaczmarz algorithm (5.1.3) consists of homogeneous linear functions

$$H = \{f_x \in L^2_{\rho_X} : x \in \mathcal{R}^n\}, \quad \text{where } f_x(\psi) := \langle x, \psi \rangle, \ \psi \in X. \tag{5.1.5}$$

Definition 5.1.2. *The sampling process associated with ρ is said to be noise-free if $\tilde{y} = f_\rho(\psi)$ almost surely. Otherwise, it is called noisy. It is said to be linear if $f_\rho \in H$ as a function in $L^2_{\rho X}$. Otherwise, it is called nonlinear.*

The least squares generalization error $\mathcal{E}(f) = \int_{\mathcal{Z}} (\tilde{y} - f(\psi))^2 d\rho$ is a well developed concept in learning theory. The assumption $E[|\tilde{y}|^2] < \infty$ on ρ ensures $f_\rho \in L^2_{\rho X}$ and $\mathcal{E}(f_\rho) < \infty$. The noise-free condition can be stated as $\mathcal{E}(f_\rho) = 0$.

It is well known that the regression function minimizes $\mathcal{E}(f)$ among all the square integral (with respect to ρ_X) functions $f \in L^2_{\rho X}$, and satisfies

$$\mathcal{E}(f) - \mathcal{E}(f_\rho) = \|f - f_\rho\|^2_{L^2_{\rho X}} = \int_X (f(\psi) - f_\rho(\psi))^2 d\rho_X. \tag{5.1.6}$$

Since the hypothesis space H is a finite dimensional subspace of $L^2_{\rho X}$, the continuous functional $\mathcal{E}(f)$ achieves a minimizer

$$f_H = \arg\min_{f \in H} \mathcal{E}(f). \tag{5.1.7}$$

From (5.1.6), we see that f_H is the best approximation of f_ρ in the subspace H. It is unique as the orthogonal projection of f_ρ onto H. It can be written as $f_H = f_{x^*}$ for some $x^* \in \mathcal{R}^n$. But such a vector x^* is not necessarily unique.

The linear condition can be stated as $f_\rho = f_H$ or $f_\rho \in H$ as functions in $L^2_{\rho X}$. So we see that the sampling process is noisy or nonlinear if and only if $\mathcal{E}(f_H) > 0$. Now we can state the first main result [107] which gives a characterization of the convergence of $\{x_t\}_t$ to some $x^* \in \mathcal{R}^n$ in expectation.

Theorem 5.1.1. *Define the sequence $\{x_t\}_t$ by (5.1.3). Assume $\mathcal{E}(f_H) > 0$. Then we have the limit $\lim_{T \to \infty} E_{z_1, \cdots, z_T} \|x_{T+1} - x^*\|^2 = 0$ for some $x^* \in \mathcal{R}^n$ if and only if*

$$\lim_{t \to \infty} \eta_t = 0 \quad and \quad \sum_{t=1}^{\infty} \eta_t = \infty. \tag{5.1.8}$$

In this case, we have

$$\sum_{t=1}^{\infty} \sqrt{E_{z_1, \cdots, z_T} \|x_{T+1} - x^*\|^2} - \infty. \tag{5.1.9}$$

Compared with the result on exponential convergence in expectation in the linear case without noise[39], the somewhat negative result (5.1.9) tells us that in the noisy setting the convergence in expectation cannot be as fast as $E_{z_1,\cdots,z_T}\|x_{T+1} - x^*\|^2 \neq O(T^{-\theta})$ for any $\theta > 2$. But for $\theta < 1$, such learning rates can be achieved by taking $\eta_t = \eta_1 t^{-\theta}$, as shown in the following second main result, to be proved in [107].

Theorem 5.1.2. *Let $\eta_t = \eta_1 t^{-\theta}$ for some $\theta \in (0,1]$ and $\eta_1 \in (0,1)$. Define the sequence $\{x_t\}_t$ by (5.1.3). Then for some $x^* \in \mathcal{R}^n$ we have*

$$E_{z_1,\cdots,z_T}\|x_{T+1} - x^*\|^2 \leqslant \begin{cases} \tilde{C}_0 T^{-\theta}, & \text{if } \theta < 1, \\ \tilde{C}_0 T^{-\lambda_r \eta_1}, & \text{if } \theta = 1, \end{cases} \tag{5.1.10}$$

where $\tilde{C}_0$ is a constant independent of $T \in N$ (given explicitly in the proof) and λ_r is the smallest positive eigenvalue of the covariance matrix C_{ρ_X} of the probability measure ρ_X defined by

$$C_{\rho_X} = E_{\rho_X}[\psi\psi^T] = \int_X \psi\psi^T d\rho_X. \tag{5.1.11}$$

The third main result is the following confidence-based estimate for the error which was proved in [107].

Theorem 5.1.3. *Assume that for some constant $M > 0$, $|\tilde{y}| \leqslant M$ almost surely. Let $\theta \in [1/2, 1]$, $\eta_t = \eta_1 t^{-\theta}$ with $0 < \eta_1 < \min\{1, \frac{1}{2\lambda_r}\}$, and $2 \leqslant T \in N$. Then for some $x^* \in \mathcal{R}^n$ and for any $0 < \delta < 1$, with confidence at least $1 - \delta$, we have*

$$\|x_{T+1} - x^*\|^2 \leqslant \begin{cases} \tilde{C}_1 T^{-\theta/2}(\log(\frac{4}{\delta}))^2 \log(T), & \text{when } \theta \in [1/2, 1), \\ \tilde{C}_1 T^{-\lambda_r \eta_1} \log(\frac{2}{\delta})\sqrt{\log(T)}, & \text{when } \theta = 1, \end{cases} \tag{5.1.12}$$

where $\tilde{C}_1$ is a positive constant independent of T or δ (explicitly in the proof[107]).

The last main result is about the almost sure convergence of the algorithm and its proof was given in [107].

Theorem 5.1.4. *Under the assumptions of Theorem 5.1.3, we have for any $\epsilon \in (0, 1]$, the following holds for some $x^* \in \mathcal{R}^n$:*

(A) When $1/2 \leqslant \theta < 1$, $\lim\limits_{t \to \infty} t^{\theta(1-\epsilon)/2} \|x_{t+1} - x^*\| = 0$ almost surely.

(B) When $\theta = 1$, $\lim\limits_{t \to \infty} t^{\lambda_r \eta_1 (1-\epsilon)} \|x_{t+1} - x^*\| = 0$ almost surely.

The relaxed randomized Kaczmarz algorithm defined by (5.1.3) may be rewritten as an online learning algorithm with output functions from the hypothesis space (5.1.5), and the main results stated above are new even in the online learning literature.

5.1.2 Learning Theory for Sparse RK

A randomized sparse Kaczmarz method (RSK) was considered in [86] and linear convergence was established under the consistency assumption again. In 2018, Yunwen Lei and Dingxuan Zhou[108] proposed an online learning algorithm, a general RSK, which is able to perform learning tasks by using sequentially arriving data or big data since each iteration involves only a single example. The relaxation of step sizes allows the algorithm to handle noisy data. Let $\mathcal{X}$ (the input set) be a nonempty measurable subset of $\mathcal{R}^n$ and $\mathcal{Y} = \mathcal{R}$ be the output set.

Definition 5.1.3. *Let $\{z_t := (x_t, y_t)\}_t \subset \mathcal{Z} = \mathcal{X} \times \mathcal{Y}$ be a sequence of input-output pairs. The RSK produces a sequence of vector pairs $\{(w_t, v_t)\}_{t \in N}$ defined iteratively with the initial pair $w_1 = v_1 = 0$ by*

$$\begin{cases} v_{t+1} = & v_t - \eta_t(\langle w_t, x_t \rangle - y_t)x_t, \\ w_{t+1} = & S_\lambda(v_{t+1}), \end{cases} \tag{5.1.13}$$

where $\{\eta_t\}$ is a sequence of positive relaxation parameters or step sizes and $S_\lambda : \mathcal{R}^n \to \mathcal{R}^n$ is the soft-thresholding operator defined componentwisely in terms of the soft-thresholding function $S_\lambda : \mathcal{R} \to \mathcal{R}$ given by $S_\lambda(v) := sgn(v) \max(|v| - \lambda, 0)$. Here $sgn(a)$ is the sign of $a \in \mathcal{R}$.

This algorithm is an online version of the linearized Bregman iteration (3.1.17) modified with a step size sequence $\{\eta_t\}_t$. It is more general than the sparse Kaczmarz algorithm (3.3.1) considered in [84] which is a special case in the sense that the input set $\mathcal{X} = \{\frac{a_r}{\|a_r\|}\}_{r=1}^{m}$ takes the special choice consisting of the normalized row vectors of A and the step size sequence takes the special constant sequence $\eta_t \equiv 1$. If $\lambda = 0$, then our algorithm recovers the randomized Kaczmarz algorithm studied in [107].

To state the limit of the vector sequence $\{w_t\}_{t \in N}$ defined by (5.1.13), we denote by $\mathcal{Z} = (\mathcal{X}, \mathcal{Y})$ a random sample drawn from a Borel probability measure ρ on $\mathcal{Z}$ and by $C_{\rho \mathcal{X}} = E_{\mathcal{Z}}[\mathcal{X} \mathcal{X}^T]$ the covariance matrix of the marginal distribution $\rho_{\mathcal{X}}$ of ρ on $\mathcal{X}$, where $E_{\mathcal{Z}}$ is the expectation with respect to $\mathcal{Z}$. Then the linear equation $C_{\rho \mathcal{X}} w = E_{\mathcal{Z}}[\mathcal{X} \mathcal{Y}^T]$ is consistent and we denote its solution set as

$$\mathcal{W}^* := \{w \in \mathcal{R}^n : C_{\rho \mathcal{X}} w = E_{\mathcal{Z}}[\mathcal{X} \mathcal{Y}]\}. \tag{5.1.14}$$

The target vector (the limit of $\{w_t\}$) is now defined by

$$w^* = \arg \min_{w \in \mathcal{W}^*} \Psi(w) \equiv \arg \min_{w \in \mathcal{W}^*} \lambda \|w\|_1 + \frac{1}{2} \|w\|_2^2. \tag{5.1.15}$$

Main Results

Assume that $\mathcal{X} \subseteq \{x \in \mathcal{R}^n : \|x\| \leqslant R\}$ for some $R > 0$ and that the target vector w^* defined by (5.1.15) is not the zero vector. Let $I = \{i \in [n] : w^*(i) \neq 0\}$ be the support of w^* and we denote by $w(I) = (w(i))_{i \in I}$ the restriction of $w \in \mathcal{R}^n$ onto the index set $\mathcal{I}$ (see [108] for the proofs of the following three theorems).

Theorem 5.1.5. . *Let $\{(wt, vt)\}_t \in N$ be the sequence generated by (5.1.13) and w^* defined by (5.1.15). Assume*

$$\inf_{w \in \mathcal{R}^n} E_{\mathcal{Z}}[\|(\langle w, \mathcal{X} \rangle - \mathcal{Y}) \mathcal{X}(I)\|] > 0. \tag{5.1.16}$$

Then $\lim_{T \to \infty} E_{z_1, \cdots, z_{T-1}}[\|w_T - w^*\|^2] = 0$ if and only if the step size sequence satisfies

$$\lim_{t \to \infty} \eta_t = 0 \quad and \quad \sum_{t=1}^{\infty} \eta_t = \infty. \tag{5.1.17}$$

In this case, we have

$$\sum_{T=1}^{\infty} \sqrt{E_{z_1,\cdots,z_{T-1}}[\|w_T - w^*\|^2]} = \infty. \tag{5.1.18}$$

Theorem 5.1.6. . *Let $\{(wt, vt)\}_t \in N$ be the sequence generated by (5.1.13) and w^* defined by (5.1.15). If the step size sequence satisfies*

$$\sum_{t=1}^{\infty} \eta_t = \infty \quad and \quad \sum_{t=1}^{\infty} \eta_t^2 < \infty, \tag{5.1.19}$$

then we have $\lim_{t\to\infty} \|w_t - w^*\|^2 = 0$ almost surely.

Condition (5.1.19) also appears in the literature to study the almost sure convergence of stochastic gradient descent algorithms[109]. It was commonly used in investigating online learning algorithms (e.g.,[110],[111]). The second part of this condition implies $\lim_{t\to\infty} \eta_t = 0$.

Theorem 5.1.7. . *Let $\{(wt, vt)\}_t \in N$ be the sequence generated by (5.1.13) and w^* defined by (5.1.15).*
(a) If we take the step size sequence as $\eta_t = \eta_1 t^{-\theta}$ with $0 < \eta_1 \leqslant (2R^2)^{-1}$ and $0 < \theta < 1$, then there exists a constant C_1 independent of T such that

$$E_{z_1,\cdots,z_T}[\|w_{T+1} - w^*\|^2] \leqslant C_1 T^{-\theta}, \quad \forall T \in N. \tag{5.1.20}$$

(b) There exist constants $\tilde{a} > 0$ and $C_2 > 0$ independent of T such that with the step size sequence $\eta_t = \frac{2}{(t+1)\tilde{a}}$,

$$E_{z_1,\cdots,z_T}[\|w_{T+1} - w^*\|^2] \leqslant C_2 T^{-1}, \quad \forall T \geqslant 4R^2 \tilde{a}^{-1}. \tag{5.1.21}$$

5.2 Kaczmarz Methods for Linear Inequalities

Iterated projection algorithms share important characteristics with coordinate descent algorithms and indeed may in some sense be considered dual to each other (Luo and Tseng [112]). In[20], Leventhal and Lewis generalized the RK algorithm and convergence result of Strohmer and Vershynin[39] to systems of linear inequalities of the form

$$\begin{cases} a_i^T x \leqslant b_i & (i \in I_\leqslant) \\ a_i^T x = b_i & (i \in I_=), \end{cases} \tag{5.2.1}$$

where the disjoint index sets $I_\leqslant$ and $I_=$ partition the set $[m]$. First, given a vector $x \in \mathcal{R}^n$, define the vector x^+ by $(x^+)_i = \max\{x_i, 0\}$. Then a starting point for this subject is a result by Hoffman.

Theorem 5.2.1. (*Hoffman[113]*). *For any right-hand side vector $b \in \mathcal{R}^m$, let S_b be the set of feasible solutions of the linear system (5.2.1). Then there exists a constant L, independent of b, with the following property:*

$$x \in \mathcal{R}^n \text{ and } S_b \neq \emptyset \ \Rightarrow \ d(x, S_b) \leqslant L\|e(Ax - b)\|, \tag{5.2.2}$$

where the function $e : \mathcal{R}^m \to \mathcal{R}^m$ is defined by

$$e(y)_i = \begin{cases} y_i^+ & (i \in I_\leqslant) \\ y_i & (i \in I_=). \end{cases}$$

In the above result, each component of the vector $e(Ax-b)$ indicates the error in the corresponding inequality or equation. In particular $e(Ax-b) = 0$ if and only if $x \in S_b$. Thus Hoffman's result provides a linear bound for the distance from a trial point x to the feasible region in terms of the size of the a posteriori error associated with x.

We call the minimum constant L such that property (5.2.2) holds, the Hoffman constant for the system (5.2.1). Several authors gave geometric or algebraic meaning to this constant, or exact expressions for it, including Güler et al.[114], Ng and Zheng[115], Li[116], Ho and Tuncel[117], Zhang[118], and the survey of Pang[119]. In the case of linear equations (that is, $I_\leqslant = \emptyset$), an easy calculation using the singular value decomposition shows that the Hoffman constant is just the reciprocal of the smallest nonzero singular value of the matrix A, and hence equals $\|A^{-1}\|$ when A has full column rank.

For the problem of finding a solution to a system of linear inequalities, we consider a randomized algorithm generalizing the randomized iterated projections algorithm.

Algorithm 5.2.1 Randomized Iterated Projections Algorithm for Inequalities (RIPKAI)

1: **procedure** $(A,\ b,\ x^{(0)},\ \mathcal{I}_{\leqslant},\ \mathcal{I}_{=})$

2:     **for** $k = 0$ to K **do**

3:         Select $i_k \in [m]$ with probability $\Pr(\text{row}{=}i_k){=}\frac{\|a_{i_k}\|}{\|A\|_F^2}$

4:         Set $\beta_k = \begin{cases} a_{i_k}^T x^{(k)} - b_{i_k}, & i_k \in \mathcal{I}_{=} \\ (a_{i_k}^T x^{(k)} - b_{i_k})^+, & i_k \in \mathcal{I}_{\leqslant} \end{cases}$

5:         Set $x^{(k+1)} = x^{(k)} + \frac{\beta_k}{\|a_{i_k}\|_2^2} a_{i_k}$

6:     **end for**

7:     Output $x^{(K+1)}$.

8: **end procedure**

In the above algorithm, notice that $\beta_k = e(Ax^{(k)} - b)_{i_k}$ and that $x^{(k+1)}$ is just the orthogonal projection onto the halfspace or hyperplane defined by the constraint with index i_k. We can now generalize Theorem 2.1.1 as follows.

Theorem 5.2.2. *Suppose the system (5.2.1) has nonempty feasible region $\mathcal{S}$. Then the randomized iterated projections algorithm for inequalities (RIP-KAI) converges linearly in expectation: for each iteration $j = 0, 1, 2, \cdots,$*

$$E[d(x^{(k+1)}, \mathcal{S})^2 | x^{(k)}] \leqslant \left(1 - \frac{1}{L^2 \|A\|_F^2}\right) d(x^{(k)}, \mathcal{S})^2,$$

where L is the Hoffman constant.

Proof. Note that if the index i_k is chosen during iteration k, then it follows that

$$
\begin{aligned}
&\|x^{(k+1)} - P_{\mathcal{S}}(x^{(k+1)})\|^2 \leqslant \|x^{(k+1)} - P_{\mathcal{S}}(x^{(k)})\|^2 \\
={}& \left\|x^{(k)} - \frac{e(Ax^{(k)}-b)_{i_k}}{\|a_{i_k}\|^2} a_{i_k} - P_{\mathcal{S}}(x^{(k)})\right\|^2 \\
={}& \|x^{(k)} - P_{\mathcal{S}}(x^{(k)})\|^2 + \frac{e(Ax^{(k)}-b)_{i_k}^2}{\|a_{i_k}\|^2} - 2\frac{e(Ax^{(k)}-b)_{i_k}}{\|a_{i_k}\|^2} a_{i_k}^T(x^{(k)} - P_{\mathcal{S}}(x^{(k)})).
\end{aligned}
$$

Note $P_{\mathcal{S}}(x^{(k)}) \in \mathcal{S}$. Hence if $i_k \in \mathcal{I}_\leqslant$, then $a_{i_k}^T P_{\mathcal{S}}(x^{(k)}) \leqslant b_{i_k}$, and $e(Ax^{(k)} - b)_{i_k}^2 \geqslant 0$, so

$$
e(Ax^{(k)}-b)_{i_k} a_{i_k}^T(x^{(k)}-P_{\mathcal{S}}(x^{(k)})) = e(Ax^{(k)}-b)_{i_k}(a_{i_k}^T x^{(k)}-b_{i_k}) = e(Ax^{(k)}-b)_{i_k}^2.
$$

Putting these two cases together with the previous inequality shows

$$
d(x^{(k+1)}, \mathcal{S})^2 \leqslant d(x^{(k)}, \mathcal{S})^2 - \frac{e(Ax^{(k)} - b)_{i_k}^2}{\|a_{i_k}\|^2}.
$$

Taking the expectation with respect to the specified probability distribution, it follows that

$$
E[d(x^{(k+1)}, \mathcal{S})^2 | x^{(k)}] \leqslant d(x^{(k)}, \mathcal{S})^2 - \frac{\|e(Ax^{(k)} - b)\|^2}{\|A\|_F^2}
$$

and the result now follows from the Hoffman bound.

Since Hoffman's bound is not independent of the scaling of the matrix A, it is not surprising that a normalizing constant such as the $\|A\|_F^2$ term appears in the result.

5.3   Other Issues

There are many applications of the Kaczmarz algorithm that have not been summarized in scientific and engineering calculation. Such as matrix factorization[120], solution of tensor linear systems of equation[121,122], quantum algorithm[123], solving quadratic equations[124], complementarity problem[125], phase retrieval algorithm[126], matrix equation[127,128] and other applications.

Bibliography

[1] S. Kaczmarz, Angenäherte auflösung von systemen linearer gleichungen, Bulletin International de l'Académie Polonaise des Sciences et des Lettres 35 (1937) 355–357.

[2] S. Kaczmarz, Approximate solution of systems of linear equations, International Journal of Control 57 (1993) 1269–1271.

[3] G. Cimmino, Calcolo approssimato per le soluzioni dei sistemi di equazioni lineari, La Ric. Sci., XVI, Ser. II, Anno IX 1 (1938) 326–333.

[4] R. Gordon, R. Bender, G. T. Herman, Algebraic Reconstruction Techniques (ART) for three-dimensional electron microscopy and X-ray photography, Journal of Theoretical Biology 29 (1970) 471–481.

[5] K. Tanabe, Projection method for solving a singular system of linear equations and its applications, Numerische Mathematik 17 (1971) 203–214.

[6] C. Popa, Convergence rates for Kaczmarz-type algorithms, Numerical Algorithm 79 (2018) 1–17.

[7] P. P. B. Eggermont, G. T. Herman, A. Lent, Iterative algorithms for large partitioned linear systems with applications to image reconstruction, Linear Algebra and Its Applications 40 (1981) 37–67.

[8] P. P. B. Eggermont, G. T. Herman, A. Lent, Two algorithms related to the method of steepest descent, SIAM Journal On Numerical Analysis 4 (1967) 109–118.

[9] P. C. Hansen, M. Saxild-Hansen, AIR tools —A MATLAB package of algebraic iterative reconstruction methods, Journal of Computational and Applied Mathematics 236 (2012) 2167–2178.

[10] P. Gilbert, Iterative methods for the reconstruction of three-dimensional objects from projection, Journal of Theoretical Biology 36 (1972) 105–117.

[11] A. Dines Kris, R. Jeffrey Lytle, Computerized geophysical tomography, proceedings of the IEEE, Journal of Japan Society for Fuzzy Theory and Systems 67 (1979) 1065–1073.

[12] B. Hager, R. Clayton, M. Richards, R. Comer, A. Dziewonsky, Lower mantle heterogeneity, dynamic typography and the geoid, Nature 313 (1985) 541–545.

[13] Y. Censor, T. Elfving, Block-iterative algorithms with diagonally s-caled oblique projections for the linear feasibility problem, SIAM Journal on Matrix Analysis and Applications 24 (2002) 40–58.

[14] G. Qu, C. Wang, M. Jiang, Necessary and sufficient convergence conditions for algebraic image reconstruction algorithms, IEEE Transactions on Image Processing A Publication of the IEEE Signal Processing Society 18 (2009) 435–440.

[15] Y. Censor, T. Elfving, G. T. Herman, T. Nikazad, On diagonally relaxed orthogonal projection methods, SIAM Journal on Scientific Computing 30 (2007) 473–504.

[16] A. H. Andersen, A. C. Kak, Simultaneous algebraic reconstruction

technique (SART): a superior implementation of the art algorithm, Ultrasonic Imaging 6 (1984) 81–94.

[17] M. Jiang, G. Wang, Convergence of the simultaneous algebraic reconstruction technique (SART), IEEE Transactions on Image Processing A Publication of the IEEE Signal Processing Society 12 (2003) 957–961.

[18] M. Yan, Convergence analysis of SART: optimization and statistics, International Journal of Computer Mathematics 90(1) (2013) 30–47.

[19] G. H. Golub, C. F. Van Loan, Matrix computations, 4th edition, The Johns Hopkins University Press, Baltimore, Maryland. (2013).

[20] D. Leventhal, A. S. Lewis, Randomized methods for linear constraints: Convergence rates and conditioning, Mathematics of Operations Research 35 (2010) 641–654.

[21] S. Kayalar, H. L. Weinert, Oblique projections: formulas, algorithms, and error bounds, Mathematics of Control Signals and Systems 2 (1989) 33–45.

[22] F. J. Murray, On complementary manifolds and projections in l_p and l_p, Transactions of the American Mathematical 41 (1937) 138–152.

[23] E. R. Lorch, On a calculus of operators in reflexive vector spaces, Transactions of the American Mathematical 45 (1939) 217–234.

[24] R. T. Behrens, L. L. Scharf, Signal processing applications of oblique projections, IEEE Transactions on Signal Processing 42 (1994) 1413–1424.

[25] M. Arioli, I. Duff, J. Noailles, D. Ruiz, A block projection method for sparse matrices, SIAM Journal on Scientific and Statistical Computing 13 (1992) 47–70.

[26] Y. Censor, D. Gordon, R. Gordon, Component averaging: An efficient iterative parallel algorithm for large and sparse unstructured problems, Parallel Computing 27 (2001) 777–808.

[27] H. H. Bauschke, J. M. Borwein, On projection algorithms for solving convex feasibility problems, SIAM Review 38 (1996) 367–426.

[28] C. Popa, Projection algorithms - classical results and developments, Lap Lambert Academic Publishing, 2012.

[29] D. A. Lorenz, S. Rose, F. Schöpfer, The randomized Kaczmarz method with mismatched adjoint, BIT Numerical Mathematics 58 (2018) 1079–1098.

[30] D. Needell, R. Ward, Two-subspace projection method for coherent overdetermined systems, Journal of Fourier Analysis and Applications 19 (2013) 256–269.

[31] W. Li, Q. Wang, W. Bao, L. Liu, On Kaczmarz method with oblique projection for solving large overdetermined linear systems, arXiv e-prints (2021-06-25) DOI:arxiv–2106.13368.

[32] W. Wu, On two-subspace randomized extended Kaczmarz method for solving large linear least-squares problems, Numerical Algorithms (2021) https://doi.org/10.1007/s11075–021–01104–x.

[33] T. Elfving, Block-iterative methods for consistent and inconsistent linear equations, Numerische Mathematik 35 (1980) 1–12.

[34] Y. Q. Niu, B. Zheng, Paving the randomized Gauss-Seidel method, Applied Mathematics Letters 104 (2020) 106–294.

[35] J. Nutini, B. Sepehry, I. Laradji, M. Schmidt, H. Koepke, A. Virani, Convergence rates for greedy Kaczmarz algorithms, and faster randomized Kaczmarz rules using the orthogonality graph, Proceedings

of the Thirty-Second Conference on Uncertainty in Artificial Intelligence, Jersey City, New Jersey, USA, (2016) 547–556.

[36] Z. Bai, W. Wenting, On greedy randomized Kaczmarz method for solving large sparse linear systems, SIAM Journal on Scientific Computing 40 (2018) A592–A606.

[37] H. Li, Y. Zhang, Greedy block Gauss-Seidel methods for solving large linear least squares problem, arXiv:2004.02476 (2020).

[38] Z. Bai, W. Wu, On greedy randomized coordinate descent methods for solving large linear least-squares problems, BIT Numerical Mathematics 26 (2019) 1–15.

[39] T. Strohmer, R. Vershynin, A randomized Kaczmarz algorithm with exponential convergence, Journal of Fourier Analysis and Applications 15 (2009) 262–278.

[40] A. Zouzias, N. M. Freris, Randomized extended Kaczmarz for solving least squares, Siam Journal on Matrix Analysis and Applications 34(2) (2013) 773–793.

[41] Y. Censor, G. T. Herman, M. Jiang, A note on the behavior of the randomized Kaczmarz algorithm of Strohmer and Vershynin, Journal of Fourier Analysis and Applications 15 (2009) 431–436.

[42] T. Strohmer, R. Vershynin, Comments on the randomized Kaczmarz method, Journal of Fourier Analysis and Applications 15 (2009) 437–440.

[43] J. Guo, W. Li, The randomized Kaczmarz method with a new random selection rule, Numerical Mathematics (Theory, Methods and Applications) 40 (2018) 65–75.

[44] Y. Guan, W. Li, T. Qiao, L. Xing, A note on convergence rate of randomized Kaczmarz method, Calcolo 57 (2020) 26.

[45] S. Steinerberger, A weighted randomized Kaczmarz method for solving linear systems, Mathematics of Computation 332 (2021) 2815–2826.

[46] A. Ma, D. Needell, A. Ramdas, Convergence properties of the randomized extended Gauss-Seidel and Kaczmarz methods, SIAM Journal on Matrix Analysis and Applications 36 (2015) 1590–1604.

[47] D. Needell, Randomized Kaczmarz solver for noisy linear systems, BIT Numerical Mathematics 50 (2010) 395–403.

[48] K. Du, Tight upper bounds for the convergence of the randomized extended Kaczmarz and Gauss-Seidel algorithms, Numerical Linear Algebra with Applications 26 (2019) 2233.

[49] A. Hefny, D. Needell, A. Ramdas, Rows versus columns: randomized Kaczmarz or Gauss-Seidel for ridge regression, SIAM Journal on Scientific Computing 39 (2017) S528–S542.

[50] F. Wang, W. Li, W. Bao, Z. Lv, Gauss-Seidel method with oblique direction, Results in Applied Mathematics 12 (2021) 100180.

[51] F. Wang, W. Li, W. Bao, L. Liu, Greedy randomized and maximal weighted residual Kaczmarz methods with oblique projection, Electronic Research Archive 30(4) (2022) 1158–1186.

[52] L. Wei, W. Li, F. Wang, Two Gauss-Seidel methods with oblique direction for solving large-scale least squares problems, Journal on Numerical Methods and Computer Applications In Press (2022).

[53] S. F. Mccormick, The methods of Kaczmarz and row orthogonalization for solving linear equations and least squares problems in Hilbert space, Indiana University Mathematical Journal 26 (1977) 1137–1150.

[54] K. Du, H. Gao, A new theoretical estimate for the convergence rate of the maximal residual Kaczmarz algorithm, Numerical Mathmatica-Theory Methods and Applications 12 (2019) 627–639.

[55] Z. Bai, W. Wu, On partially randomized extended Kaczmarz method for solving large sparse overdetermined inconsistent linear systems, Linear Algebra and its Applications 578 (2019) 225–250.

[56] W. Bao, Z. Lv, F. Zhang, W. Li, A class of residual-based extended kaczmarz methods for solving inconsistent linear systems, Journal of Computational and Applied Mathematics 416 (2022) 114529.

[57] R. K. Meanys, A matrix inequality, SIAM Journal on Numerical Analysis 6 (1969) 104–107.

[58] C. Popa, Correction to: convergence rates for Kaczmarz-type algorithms, Numerical Algorithms 1 (2019) 1–4.

[59] Y. Jiang, G. Wu, L. Jiang, A Kaczmarz method with simple random sampling for solving large linear systems, arXiv.org (2020) 1–26.

[60] Z. Lv, W. Bao, W. Li, F. Wang, G. Wu, On extended Kaczmarz methods with random sampling and maximum-distance for solving large inconsistent linear systems, Results in Applied Mathematics 13 (2022) 100240.

[61] Y. Nesterov, Efficiency of coordinate descent methods on huge-scale optimization problems, SIAM Journal on Optimization 22 (2012) 341–362.

[62] Y. T. Lee, A. Sidford, Efficient accelerated coordinate descent methods and faster algorithms for solving linear systems, IEEE Symposium on Foundations of Computer Science 54 (2013) 147–156.

[63] D. Needell, J. A. Tropp, Paved with good intentions: analysis of a randomized block Kaczmarz method, Linear Algebra and its Applications 441 (2014) 199–221.

[64] D. Needell, R. Zhao, A. Zouzias, Randomized block Kaczmarz method with projection for solving least squares, Linear Algebra and Its Applications 484 (2015) 322–343.

[65] R. M. Gower, P. Richt/arik, Randomized iterative methods for linear systems, SIAM Journal on Matrix Analysis and Applications 36 (2015) 1660–1690.

[66] I. Necoara, Faster randomized block Kaczmarz algorithms, SIAM Journal on Matrix Analysis and Applications 40 (2019) 1425–1452.

[67] W. Wu, Paving the randomized Gauss-Seidel method, BSc Thesis, Scripps College, Claremont, California (2017).

[68] K. Du, X. H. Sun, Pseudoinverse-free randomized block iterative algorithms for consistent and inconsistent linear systems, arXiv:2011.10353v2 (2020).

[69] M. Razaviyayn, M. Hong, N. Reyhanian, L. Z.Q., A linearly convergent doubly stochastic Gauss-Seidel algorithm for solving linear equations and a certain class of over-parameterized optimization problems, Mathematical Programming 176 (2019) 465–496.

[70] Y. F. Lou, S. Osher, J. Xin, Computational aspects of constrained l_1-l_2 minimization for compressive sensing, Advances in Intelligent Systems and Computing 359 (2015) 169–180.

[71] E. J. Candes, J. Romberg, T. Tao, Robust uncertainty principles: Exact signal reconstruction from highly incomplete frequency information, IEEE Transactions on Information Theory 52 (2006) 489–509.

[72] W. T. Yin, Analysis and generalizations of the linearized Bregman method, SIAM Journal on Imaging Sciences 3(4) (2010) 856–877.

[73] S. Osher, M. Burger, D. Goldfarb, J. Xu, W. Yin, An iterative regularization method for total variation-based image restoration, SIAM Journal on Multiscale Modeling and Simulation 4 (2005) 460–489.

[74] W. Yin, S. Osher, D. Goldfarb, J. Darbon, Bregman iterative algorithms for l_1−minimization with applications to compressed sensing, SIAM Journal on Imaging Sciences 1 (2008) 143–168.

[75] S. Osher, Y. Mao, B. Dong, W. T. Yin, Fast linearized Bregman iteration for compressive sensing and sparse denoising, Communications in Mathematical Sciences 8 (2010) 93–111.

[76] J. F. Cai, S. Osher, Z. Shen, Linearized Bregman iterations for compressed sensing, Mathematics of Computation 78 (2009) 1515–1536.

[77] J. F. Cai, S. Osher, Z. Shen, Convergence of the linearized Bregman iteration for l_1−norm minimization, Mathematics of Computation 78 (2009) 2127–2136.

[78] J. S. Zeng, S. B. Lin, Z. B. Xu, Sparse solution of underdetermined linear equations via adaptively iterative thresholding, Signal Processing 97 (2014) 152–161.

[79] Z. S. Zhang, Y. C. Yu, S. M. Zhao, Iterative hard thresholding based on randomized Kaczmarz method, Circuits, Systems, and Signal Processing 34 (2015) 2065–2075.

[80] T. Blumensath, M. E. Davies, Iterative thresholding for sparse approximations, Journal of Fourier Analysis and Applications 14(5) (2008) 629–654.

[81] T. Blumensath, M. E. Davies, Iterative hard thresholding for compressed sensing, Applied and Computational Harmonic Analysis 27(3) (2009) 265–274.

[82] T. Blumensath, M. E. Davies, Normalised iterative hard thresholding: guaranteed stability and performance, IEEE Journal of Selected Topics in Signal Processing 4(2) (2010) 298–309.

[83] D. A. Lorenz, F. Schöpfer, S. Wenger, The linearized Bregman method via split feasibility problems: analysis and generalizations, SIAM Journal on Imaging Sciences 7(2) (2014) 1237–1262.

[84] D. A. Lorenz, S. Wenger, F. Schöpfer, M. Magnor, A sparse Kaczmarz solver and a linearized Bregman method for online compressed sensing, IEEE International Conference on Image Processing (2014) 1347–1351.

[85] F. Schöpfer, Exact regularization of polyhedral norms, SIAM Journal on Optimization 22(4) (2012) 1206–1223.

[86] F. Schöpfer, D. A. Lorenz, Linear convergence of the randomized sparse Kaczmarz method, Mathematical Programming 173 (2019) 506–536.

[87] F. Schöpfer, Linear convergence of descent methods for the unconstrained minimization of restricted strongly convex functions, SIAM Journal on Optimization 26(3) (2016) 1883–1911.

[88] R. T. Rockafellar, R. J. B. Wets, Variational Analysis, Springer, Berlin, 2004.

[89] M. Hanke, A. Neubauer, O. Scherzer, A convergence analysis of the Landweber iteration for nonlinear ill-posed problems, Numerische Mathematik 72 (1995) 21–37.

[90] R. Kowar, O. Scherzer, Convergence analysis of a Landweber-Kaczmarz method for solving nonlinear ill-posed problems, Journal of Inverse and Ill-Posed Problems (book series) 23 (2002) 69–90.

[91] K. Du, H. Gao, On the convergence of stochastic gradient descent for nonlinear ill-posed problems, SIAM Journal on Optimization 30(2) (2020) 1421–1450.

[92] H. Robbins, S. Monro, A stochastic approximation method, The Annals of Mathematical Statistics 22 (1951) 400–407.

[93] H. J. Kushner, G. G. Yin, Stochastic Approximation and Recursive Algorithms and Applications, Springer-Verlag, New York, 2003.

[94] K. Chen, Q. Li, J. G. Liu, Online learning in optical tomography: A stochastic approach, Inverse Problems 34 (2018) 075010.

[95] Y. Jiao, B. Jin, X. Lu, Preasymptotic convergence of randomized kaczmarz method, Inverse Problems 33 (2017) 125012.

[96] B. Jin, X. Lu, On the regularizing property of stochastic gradient descent, Inverse Problems 35 (2019) 015004.

[97] M. Haltmeier, A. Leitao, O. Scherzer, Kaczmarz methods for regularizing nonlinear ill-posed equations. I. convergence analysis, Inverse Problems and Imaging 1 (2007) 289–298.

[98] A. De Cezaro, M. Haltmeier, A. Leitao, O. Scherzer, On steepest-descent-Kaczmarz methods for regularizing systems of nonlinear ill-posed equations, Applied Mathematics and Computation 202 (2008) 596–607.

[99] M. Burger, B. Kaltenbacher, Regularizing Newton-Kaczmarz methods for nonlinear ill-posed problems, SIAM Journal on Numerical Analysis 44 (2006) 153–182.

[100] B. Kaltenbacher, Some Newton type methods for the regularization of nonlinear ill-posed problems, Inverse Problems 13 (1997) 729–753.

[101] A. B. Bakushinskii, The problem of the convergence of the iteratively regularized Gauss-Newton method, Computational Mathematics and Mathematical Physics 32 (1992) 1353–1359.

[102] A. B. Bakushinskii, M. Y. Kokurin, Iterative Methods for Approximate Solution of Inverse Problems, Springer, Dordrecht, 2004.

[103] J. Baumeister, B. Kaltenbacher, A. Leitao, On Levenberg-Marquardt-Kaczmarz iterative methods for solving systems of nonlinear ill-posed equations, Inverse Problems and Imaging 4 (2010) 335–350.

[104] H. Li, M. Haltmeier, The averaged Kaczmarz iteration for solving inverse problems, SIAM Journal on Imaging Sciences 11 (2018) 618–642.

[105] Q. Wang, W. Li, W. Bao, X. Gao, Nonlinear Kaczmarz algorithms and their convergence, Journal of Computational and Applied Mathematics 399(1) (2022) 113720.

[106] X. Chen, A. Powell, Almost sure convergence for the Kaczmarz algorithm with random measurements, Journal of Fourier Analysis and Applications 18 (2012) 1195–1214.

[107] J. Lin, D. Zhou, Learning theory of randomized Kaczmarz algorithm, The Journal of Machine Learning Research 16 (2015) 3341–3365.

[108] Y. Lei, D. Zhou, Learning theory of randomized sparse Kaczmarz algorithm, SIAM Journal on Imaging Sciences 11 (2018) 547–574.

[109] L. Bottou, On-line learning and stochastic approximations, Cambridge University Press, New York, NY, 1999.

[110] P. Tarres, Y. Yao, Online learning as stochastic approximation of regularization paths: Optimality and almost-sure convergence, IEEE Transactions on Information Theory 60 (2014) 5716–5735.

[111] Y. Y., D. X. Zhou, Online regularized classification algorithms, IEEE Transactions on Information Theory 52 (2006) 4775–4788.

[112] Z. Luo, P. Tseng, On the convergence of the coordinate descent method for convex differentiable minimization, Journal of Optimization Theory and Applications 72 (1992) 7–35.

[113] A. Hoffman, On approximate solutions of systems of linear inequalities, Journal of Research of the National Bureau of Standards 49 (1952) 263–265.

[114] O. Güler, A. J. Hoffman, U. Rothblum, Approximations to solutions to systems of linear inequalities, SIAM Journal on Matrix Analysis and Applications 16 (1995) 688–696.

[115] K. F. Ng, X. Y. Zheng, Hoffman's least error bounds for systems of linear inequalities, Journal of Global Optimization 30 (2004) 391–403.

[116] W. Li, The sharp Lipschitz constants for feasible and optimal solutions of a perturbed linear program, Linear Algebra and its Applications 187 (1993) 15–40.

[117] J. C. K. Ho, L. Tuncel, Reconciliation of various complexity and condition measures for linear programming problems and a generalization of Tardos' theorem, Foundations of Computational Mathematics, 2000.

[118] S. Zhang, Global error bounds for convex conic problems, SIAM Journal on Optimization 10 (2000) 836–851.

[119] J. S. Pang, Error bounds in mathematical programming, Mathematical Programming 79 (1997) 299–332.

[120] E. Chau, On application of block Kaczmarz methods in matrix factorization, arXiv:2010.10635v1 (2020).

[121] A. Ma, D. Molitor, Randomized Kaczmarz for tensor linear systems, BIT Numerical Mathematics 62 (2022) 171–194.

[122] X. Chen, J. Qin, Regularized Kaczmarz algorithms for tensor recovery, SIAM Journal on Imaging Sciences 14(4) (2021).

[123] X. Liu, J. Wang, M. Li, S. Shen, W. Li, S. Fei, Quantum relaxed row and column iteration methods based on block-encoding, Quantum Information Processing 21 (2022) 230.

[124] Y. Chi, Y. Lu, Kaczmarz method for solving quadratic equations, IEEE Signal Processing Letters 23(9) (2016) 1183–1187.

[125] X. Wang, M. Che, Y. Wei, Randomized Kaczmarz methods for tensor complementarity problems, Computational Optimization and Applications 82 (2022) 595–615.

[126] Y. Xian, H. Liu, X. Tai, Y. Wang, Randomized Kaczmarz method for single particle x-ray image phase retrieval, arXiv:2207.04736v1 (2022).

[127] N. Wu, C. Liu, Q. Zuo, On the Kaczmarz methods based on relaxed greedy selection for solving matrix equation $AXB = C$, Journal of Computational and Applied Mathematics 413 (2022) 114374.

[128] N. Yuqi, B. Zheng, On global randomized block Kaczmarz algorithm for solving large-scale matrix equations, arXiv:2204.13920v1 (2022).